**Networks for
Research and Education**

The MIT Press
Cambridge, Massachusetts
and London, England

The papers, discussions, and analysis of three seminars on computer networking conducted by EDUCOM with support from the National Science Foundation.

**Networks for
Research and Education:**

Sharing Computer
and Information Resources
Nationwide

Edited by
Martin Greenberger, director
Julius Aronofsky
James L. McKenney
William F. Massy

Copyright © 1974 by
The Massachusetts Institute of Technology

This book was printed and bound
in the United States of America.

Library of Congress Cataloging in Publication Data
Main entry under title:

Networks for research and education. *see slip*

  Contains the papers of 3 working semi-
nars, held in 1972–73, conducted by
EDUCOM and sponsored by the Na-
tional Science Foundation.
  Includes bibliographical references.
  1. Electronic data processing in research
—Congresses. 2. Electronic data processing
—Education—Congresses. 3. Data trans-
mission systems—Congresses. 4. Research—
United States—Congresses. I. Greenberger,
Martin, 1931– ed. II. Educom. III. United
States. National Science Foundation.
Q180.55.E4N47 1973 001.4'3'02854 73-18376

ISBN 0-262-07057-X

Contents

Preface

A medical researcher sits at an on-line terminal in Honolulu searching an index to the world's medical literature stored on a computer in Bethesda, Maryland, over 5,000 miles away. His request passes across a radio network of the University of Hawaii. Arriving at a centrally placed message-processing minicomputer, it is operated on, then turned over to the Hawaiian telephone company's network for transmittal across the Pacific Ocean via an international satellite. In the continental United States the request is operated on once more by a message processor of a nationwide research network, then converted and routed to a commercial time-sharing network that moves it along to the medical information system in Bethesda. By mail the request would have taken several days. By computer-communication networks it takes less than five seconds. The response to the request, a set of literature citations, starts printing out at the terminal back in Honolulu within fifteen seconds from the time the request was dispatched.

Fancy? Exaggeration? No, essentially fact. When? In five years? Not at all. Right now. Perhaps the illustration is a bit dramatic. Perhaps it is enough to note that everyday domestic usage of the on-line medical information system through the commercial time-sharing network has been doubling every six months. It might be even better to mention the countless other important uses of networks that are becoming routine in a variety of areas. The point is that interest in information and computer networks is at an all-time high. The possibilities they offer in research and education inspire hope and confidence; among them new and better computing and information services, greater efficiency in operations, broader markets, widespread access to facilities, and extensive resource sharing. But the road to these benefits is not well paved. Many knotty problems and questions have yet to be answered—or even asked.

Responding to the heightened interest in the possibilities of networks, and reflecting its own continuing interest in improving the use of new technologies in research and education, the National Science Foundation in 1972 annouced the mounting of "an expanded research program . . . to explore . . . the resource-sharing potential of a national network in support of research and education."[1] The NSF was well aware of the obstacles and uncertainties, and it knew that although shareable resources and pockets of relevant information and experience existed, many of the people who should be involved in planning were not currently informed or discussing the possibilities with one another.[2]

One of the first projects the NSF sponsored in its new program was a series of three two-day working seminars conducted by EDUCOM. The seminars, held in late 1972 and early 1973, were designed to help identify the central issues in building and operating networks on a national basis. EDUCOM has been concerned since its founding in 1964 with fostering the collaboration of colleges and universities in the use of computer and communication technologies to further the aims of higher education. It has given the subject of networks special emphasis, beginning with its July 1966 summer study in Boulder, Colorado,[3] and continuing with the open conferences it recently has been holding twice each year.[4] EDUCOM was well suited to the job of bringing people together for the needed communication, discussion, and analysis.

EDUCOM asked Martin Greenberger to direct the seminars. In view of the nature and breadth of the subject and consonant with the spirit of network sharing, he invited Julius Aronofsky, James McKenney, and William Massy to join him in forming a faculty team. The four of us worked together closely, collaborating on the planning and execution of the seminars and the preparation of this book. Henry Chauncey, president of EDUCOM, maintained a high degree of personal interest and helpful involvement throughout the project, and Carolyn Landis, secretary of EDUCOM, handled all arrangements for the seminars with great competence and aplomb. Don Aufenkamp and Edward Weiss of the National Science Foundation gave generously of their good sense and ideas, and many others helped in selecting participants, commenting on the text, and providing useful suggestions and general support. Charles Pepper assisted with the editing of the papers and workshop reports.

Planning involved a series of decisions on the scope and nature of the seminars:

• Although the concept of a national science computer network clearly merited serious consideration and discussion, we decided that it should not be the single or necessarily even the central subject of concern. We used the term *networking* to designate the more general set of activities and arrangements whereby computers and communications are used for extensive resource sharing by a large number of separate, independent organizations. We chose "networking for research and education on the national level" to be the main subject of discussion at the seminars.

• Although the nature of higher education and its current financial problems make the possibilities of networking there most attractive, we considered it important for the seminars to embrace all institutions of research and education, including those in the profit-making sector of society.

• Although we recognized that a good understanding of the technology is at the very heart of the subject under discussion, we expected that the most difficult problems that would have to be treated at the seminars would not be technical ones. We therefore resolved to strive for a mixture of highly expert technologists with social scientists, physical scientists, decision makers, and others from a variety of fields besides the computer and communication fields.[5]

• Although many of the best-known people tend to be at the largest institutions with the greatest resources, we considered it essential to make a special effort to locate knowledgeable people at smaller institutions with lesser resources. We did this by spreading as wide a net as we knew how. We asked dozens of recognized leaders to recommend persons they felt should be invited. In addition, we set aside a substantial number of places for applications requested from a broad spectrum of fields and disciplines. The selection was not easy. By the third seminar there were as many as twelve applications for every available spot.

• Finally, we decided (again in the spirit of network sharing) that each participant should have the opportunity and encouragement personally to contribute to the seminars; that they should be *working* seminars.

The format established to provide a working environment was to divide the fifty plus participants at each of the three seminars into four workshops. Each workshop would be led by one of us, and each would address a separate set of issues related to the theme of the seminar. The seminar themes were chosen to progress somewhat logically from the most general and definitional in nature to the most specific and operational:

• Seminar 1. "User Characteristics and Needs," with the four workshops organized by the way in which computing and information are used.
• Seminar 2. "Organizational Matters," with the workshops covering the topics of network management, institutional relations, user organizations, and regional computing systems.
• Seminar 3. "Operations and Funding," with workshops on computers and communications, software systems and operating procedures, applications development and user services, and network economics and funding.

At each of the seminars, a set of prepared and invited papers was presented to the group at large prior to its breaking up into workshops.

In some cases a paper was intended to provide a background or position statement for a particular workshop to help get the discussion started. Minimal effort was devoted to avoiding overlap between papers and between workshops. As a result, in more than one instance a single issue was treated several times, in several places, from several points of view. Perhaps this was *not* in the true spirit of network sharing, but we did not wish to impose restrictions that might inhibit the discussion and discourage new thoughts or differing opinions. At the beginning of the second and third seminars, we briefed participants on the highlights of what had taken place at the previous seminars, and at the end of each seminar we attempted to distill a sense of what had taken place there, again for the benefit of the group at large.

The book presents the content of the three working seminars, including edited versions of the twenty-five prepared papers and twelve workshop reports. There are four parts to the book. Parts II, III, and IV contain the edited papers and reports of seminars 1, 2, and 3, in that order. Part I, a selection and condensation of the more detailed material, contains an introductory chapter framing some of the questions considered, a chapter outlining selected highlights of issues discussed, and a chapter—perhaps the one that will be of greatest interest to most readers—summarizing the chief conclusions and recommendations of the 152 seminar participants insofar as we were able to infer them. Additional conclusions and recommendations of individuals and workshops appear throughout the papers and reports of Parts II, III, and IV.

This book is addressed to everyone with an interest in the possibilities and problems of networking for research and education, irrespective of background or occupation. It is addressed to potential users, suppliers, and especially decision makers and their staffs, including state legislators, government officials, and administrative officers of colleges, universities, and other institutions of research and education. It is intended for those who are interested in the subject but were not able to attend the seminars as well as those who were there and now wish to have a permanent record of what took place. We hope the book will make interesting reading and be useful to all these people. We recognize that at best it is only a beginning.

Communication and discussion are important first steps. That was the purpose of the working seminars. The purpose of this book is to extend the ideas explored and experiences shared in the seminars well beyond the confines of the group it was possible to include, and also beyond the time span in which the seminars took place, to help in deciding future steps.

Baltimore   Martin Greenberger, Seminar Director
April 1973   Julius Aronofsky
                                       James L. McKenney
                                       William F. Massy

References

1. "Expanded Research Program Relative to a National Science Computer Network," NSF Report 72-16, National Science Foundation, Washington, D.C., July 1972.

2. D. Don Aufenkamp and E. C. Weiss, "NSF Activities Related to a National Science Computer Network," *Proceedings of the First International Conference on Computer Communications*, Washington, D.C., October 1972, pp. 226–232.

3. George W. Brown, James G. Miller, and Thomas A. Keenan, *EDUNET—Report of the Summer Study on Information Networks*, Wiley, New York, 1967.

4. See, for example, *Networks for Higher Education*, Proceedings of the EDUCOM Spring 1972 Conference, EDUCOM, Princeton, N.J.; also, *Networks and Disciplines*, Proceedings of the EDUCOM Fall 1972 Conference, EDUCOM, Princeton, N.J.

5. A similar mixture of people was assembled in a 1969–1970 series of symposia in Washington, D.C., sponsored jointly by the Brookings Institution and The Johns Hopkins University on the allied subject of *Computers, Communications, and the Public Interest*, the name also of the book containing the papers and discussions of the symposia; Martin Greenberger, ed., The Johns Hopkins Press, Baltimore, 1971.

Overview

1 Introduction

There are, by estimate, 7,000 computers in the United States today whose primary business is processing information for research and education. Their combined annual operating budget runs into the billions of dollars, of which over $600 million is in higher education alone.[1] The programs they run and the information they use are similar in general character and often in specific detail. These computers represent a national asset of considerable importance by virtue of the magnitude and nature of the work they perform and the missions to which their work contributes. How is this important national asset organized and managed? Is its use effective in tapping the technology's potential?

One might expect the operation of these 7,000 computers to be heavily concentrated in relatively few places. One might expect that the scarce, expensive, talented people required to run really good operations would tend to prefer being close to one another in intellectually stimulating environments rather than being dispersed in the outer reaches. One might expect concentration to enlarge the clientele for the best systems (often produced by small, dedicated, innovative groups with little marketing motivation or skill) and thereby to help justify the extensive, highly demanding process of creative refinement and attentive support required to market the system, successfully match it to customer needs, and give it effective documentation and maintenance. Finally, one might expect some economies of scale to accrue from consolidating operations for certain kinds of computation and data activity.

All these factors might seem to argue, if not for concentration of computers, at least for their integrated operation. Yet the overriding pattern is quite the opposite. The 7,000 computers are widely distributed in a large number of autonomous centers and laboratories that are separately staffed and managed. Autonomy and separateness are the rule, not the exception.

To understand the reasons for the autonomy and separateness, one must review the thirty-year-old history of information processing in research and education and study how computers were introduced, developed, and marketed.[2] The computer manufacturers played a role through their sales strategies and discount policies, as did the federal government with its funding programs and restrictive rules, the research and educational organizations with their rivalries and proprietary instincts, and the minicomputers with their impressive performances and increasingly

low costs. Many say that the resulting situation is not necessarily unfavorable. The diversity it breeds has much to commend it. Some say
it may even be for the best, although others wonder. In any case, the
situation today, by and large, consists of many separate, autonomous
computer installations.

Given the commonality in the work performed by these many centers,
one might expect them to engage in a great deal of sharing of data, programs, and other computing resources. Such expectation is not borne out
by current practice. Some sharing does take place, but its range is limited,
and it is beset by problems, not the least of which are the fundamental
incompatibilities that persist in equipment, data formats, programs, and
operating systems. Except for some cooperative groups of computer users
and the customer-support initiative of computer manufacturers, few
organized sharing arrangements of any magnitude have prospered and
grown. Many have failed.[3]

Thus, along with the autonomy and separateness of the computing
operations, there goes typically a self-sufficiency in the maintenance, use,
and often even development of computing resources. The lack of sharing
is not all bad. It encourages innovation and stimulates a variety of different approaches. But self-sufficiency also can be wasteful, redundant,
and expensive, especially when it results, as it often does, from an iron
determination to maintain local control or from a heel-digging resolute-
ness not to use the other fellow's system. Of course, sometimes it is
genuinely difficult to use the other fellow's system.

Four significant technological trends are beginning to affect the present
organization of computing services:

1. Minicomputers selling for as little as $20,000 (and less) are becoming
ever more powerful and popular, allowing computer users who were
formerly customers of large computer centers to purchase and operate
their own equipment.

2. The use of computers remotely, either in a time-sharing or remote
job entry mode, is gaining in acceptance. This development frees users to
look to outside suppliers (to industry or other institutions, for example)
for the best service and price for their purposes.

3. Improvements in computer communications technology and in
data transmission and switching procedures are making it easier and less
costly to have connections between unlike computers running under
unlike operating systems across great distances. The resulting networks
are spreading and are beginning to provide what could become an important new mode, called *networking*, for sharing the resources and linking

the otherwise incompatible procedures and formats of different systems and organizations.

4. Large, cheap memories make the amassing of ever bigger information banks feasible and dependent primarily on the cost of developing and maintaining the bank, thus tending to favor a single, centralized information storage and retrieval operation over numerous dispersed operations.

Many people regard the minicomputer trend as running counter to the remote computing and network trends. Users who acquire their own computers, after all, seem less likely to require the services of a remote computer or computer network. In a different sense, however, the trends are complementary and mutually reinforcing. They all work to reduce the dependency of the user on the local center; they accord the user more options; and they increase the responsibility of the user to plan, organize, and choose the best course available. These trends do not necessarily spell the doom of the local center, but they do suggest that the traditional autonomy of the center is shifting toward the user. The functions and services performed by the center will have to adapt accordingly. The center may be called on more and more, for example, to help the user plan and choose, even when the choice is for information services outside the institution or on the user's minicomputer rather than at the center itself. Not all users may want as much autonomy and responsibility as the new situation permits, and they may look to a somewhat altered kind of center for advice, brokerage services, and general assistance.

Current financial strains are causing many institutions of research and education to pay particularly close attention to the opportunities for importing and exporting services afforded by the technological advances. Some institutions have closed down their central computer operations. Harvard University, for example, has removed the shingle from its central facility and returned its large computer to IBM (see Chapter 33). In taking this action Harvard at first planned to develop a joint computer center with MIT, but this plan gave way to an arrangement wherein Harvard essentially became a customer of MIT. Most recently, Harvard has set out to shop nationally for the best service available for each application. Other institutions may be expected to follow its example.

As an experiment, one of the routes Harvard is taking in its national shopping tour is through the computer network known as ARPANET, developed by the Advanced Research Projects Agency of the Department of Defense. ARPANET interconnects several dozen heterogeneous and independent computer centers from coast to coast using broad bandwidth lines and an innovative communication procedure referred to by its developers as "packet switching" (see Chapter 6). Two of the large computer centers on the ARPANET in California were reported to be receiving as much as 20 to 25 percent of their total revenue from customers on the network within their first year of accepting such business. The University of Illinois is said to have given up its large Burroughs computer in order to use a similar computer at one of these network centers in San Diego.

ARPANET is just one illustration of networking. The National Science Foundation over the past few years has supported the development of about thirty regional networks among colleges and universities (see Chapters 8, 30, and 34). Here the using institutions are often smaller schools with minimal computing facilities of their own. The better-known of this class of networks include the ones centered at Dartmouth College, the University of Iowa, the University of Texas, and Oregon State University. Another form of network is the Triangle Universities system (TUCC), which provides computing services to the University of North Carolina, North Carolina State University, Duke University, and other educational institutions throughout the state (see Chapter 24). The MERIT network is a resource-sharing arrangement among the University of Michigan, Michigan State University, and Wayne State University. A new network service, UNI-COLL, was organized by the University City Science Center of Philadelphia to operate a combined computation center for a number of colleges and universities in the Delaware Valley by building upon the University of Pennsylvania's computing system (see Chapters 20 and 36). A number of states—including California, Missouri, Illinois, New Jersey, New York, Minnesota, Oregon, and Florida—either already have or are developing statewide computer networks.

Networks are also being developed to widen the availability of science information services. The National Science Foundation is funding at the University of Georgia the construction of an information system with primary focus in chemistry and biology that will provide services to about thirty colleges and universities by means of the NSF-supported Georgia regional computing network. An information system under development at UCLA will eventually service the nine campuses of the University of California and will include forty rapid response terminals strategically located around the UCLA campus. Remote TV terminals at Lehigh University permit on-line users to query current articles and abstracts in a natural language, a service that Lehigh also makes available remotely to six other academic institutions in the NSF-supported Lehigh Valley Regional Computing Network.

In the library world, the Ohio College Library Center is operating an on-line shared cataloging service for forty-eight college and university libraries. This network has been extended to university libraries in New England, New York, and Pennsylvania and is about to be replicated in other regions of the country. The National Library of Medicine has established on-line search facilities to its vast medical library through the nationwide MEDLINE service, using the TYMNET communication network of the TYMSHARE Corporation (see Chapter 35). Massachusetts General Hospital and the Systems Development Corporation also provide national services via TYMNET.

In addition to TYMSHARE Corporation (see Chapter 7), a number of other commercial firms, including the General Electric Company (see Chapter 19), University Computing Company, and Computer Sciences Corporation, operate national or, in some cases, international general-purpose computer networks. Several companies such as Keydata Corporation, Data Resources, Inc., and Interactive Data Corporation use communications to furnish highly specialized on-line services and information resources to clients across the country. The over-the-counter securities market has an on-line national network for prices and quotations (NASDAQ). And the list goes on.

What will this spread of networking activity lead to? The networking advocate sees in the trend profound possibilities for improving the organization of information processing and for expanding the sharing of information and program resources. The benefits envisioned include:

- A much greater variety and richness of available resources and a more flexible intermingling of information with computing services.
- Widened availability of resources to all institutions irrespective of size, location, or financial status.
- A decreasing cost per unit of information stored or processed because of increasing economies of scale and expanded sharing.
- Payment for information processing as it is obtained, with virtual elimination of the capital costs and budgetary uncertainties currently characteristic of autonomous information and computing center operations.

In short, the advocate sees networking as leading to greater integration, resource sharing, and availability and thereby to more and better services and higher efficiency.

Is this an oversimplification, or is the advocacy of networking fundamentally sound? If indeed it is sound, what obstacles stand in the way of further network development? What new problems will networking introduce? Can they be anticipated and treated in advance? Are there really any resources worth sharing? Where are they? How are new shareable resources best developed, maintained, and distributed? With multiple vendors and many different using and supplying institutions involved, do support requirements become unmanageable? What people would benefit most from networking? Where are they located? Need they be organized? In what ways? How are suppliers made financially liable and accountable?

These are a few of the questions considered in the working seminars on which this volume reports. The following pages of Part I outline and summarize the ideas and points of view put forth in the seminar presentations and in the workshop discussions in which 152 knowledgeable, interested, and involved people from education, government, and industry took part. Chapter 2 outlines some of the main issues, including those on which no clear consensus was reached, while Chapter 3 presents points on which there was enough agreement and commonality of opinion to warrant their being called "conclusions and recommendations" of the seminars.

References

1. See John Hamblen, "Inventory of Computers in Higher Education: 1969–70," NSF Report 72-9; National Science Foundation, Washington, D.C., 1972. Also R. V. DeGrasse, "Remote Computing in Higher Education: Prospects for the Future," University of Vermont, December 1971. Projecting from figures given in those documents and subtracting the number of computers used primarily for administration, there are approximately 3,000 computers in 1973 within higher education working on research and education. The total cost of running these 3,000 computers is over $600 million per year. At the time of NSF Report 70-46 on "National Patterns of R&D Resources: 1953–1971" (December 1970), the annual national research and development expenditure in government and industry was $25 billion, excluding $3 billion for R&D in universities. Of the $25 billion, 3 percent, or $750 million went for computing. Estimating conservatively that the corresponding figure today is $800 million, a total operating cost of about $200,000 per computer would suggest 4,000 computers for R&D outside higher education, giving the estimated total of 7,000 computers.

2. Martin Greenberger, "Development of Computing in Higher Education," *The Financing and Organization of Computing in Higher Education: 1971*, Proceedings of the EDUCOM 1971 Spring Conference, EDUCOM, Princeton, N.J., 1971, pp. 1–9.

3. See, for example, John Legates, "The Lessons of EIN," *EDUCOM Bulletin*, Vol. 7, No. 2, Summer 1972, pp. 18–20.

2 Highlights of the Issues Discussed

Many participants came to the seminars expecting to spend the time discussing the desirability of establishing a National Science Computer Network and the steps that would have to be taken in moving toward this goal. Some may have come prepared to debate the issue. To their possible surprise—perhaps pleasant—the main subject for discussion turned out to be the more general topic of "networking on the national level," which includes the concept of a National Science Computer Network but is not restricted to it.

A National Science Computer Network
One rather polar way of conceiving of the concept of a national computer network is as a single integrated framework for resource sharing in which a national organization of institutions of research and education have their information and computer facilities interconnected electrically. This picture tends to evoke strong reactions—on both sides. Some believe the potential advantages of such a network for research, education, and the country at large, would more than justify the launching of a major national effort to bring it about. Others feel a concerted push by the nation toward such an objective might detract from work on regional networks and drain away crucial financial support on which this important work depends. Some even see the prospect of a national network as having the earmarks of a national boondoggle and view interest in the idea as a personal threat to their own research projects.

There were different opinions on how a national network could or should evolve. Some pointed to the need for a carefully articulated master plan. Others favored a decentralized approach that would preserve much of the autonomy of present information activities and computing centers and minimize the amount of nationwide control. Still others envisioned continued development of several national networks: some private, some public; some resulting from the interconnection of regional networks; some developed and operated by commercial firms; and some evolving directly from ARPANET and its successors.

On this issue, as on all of the issues outlined in the present chapter (in contrast to the next one), no consensus or generally agreed-upon conclusions were reached. But examining the implications of the various alternatives did help participants understand better their own positions as well as opposing points of view.

Networking Compared to Time Sharing

Descending from the grandiose idea of a single integrated national network to a more basic level, we ask what properties must a "network" have to earn that designation? Is a time-sharing system a network? A time-sharing system *does* connect multiple users (who may be geographically dispersed) to a remote source of computer and information services. It *does* employ a communications network to effect this connection. It *does* depend on strict compliance to rules and conventions for its successful operation. These are properties of computer networks. But a time-sharing system *also* does (in its primary incarnation) involve only one central computing apparatus, manufactured by one vendor, operating under one program executive, managed and accountable to one organization. In short, it is a single supplier. For this reason, except as a limiting case, many observers probably would not regard a time-sharing system by itself as an interesting example of a computer network. It is, of course, a fundamental component of computer networks.

As tenuous as these distinctions may seem to some, it is interesting to distinguish between a single-access computer (SAC), a time-sharing system (TSS), and a computer network (NET). A TSS, with its one supplier and many simultaneous users, stands like an intermediate stage (or geometric mean) between a SAC, with its one supplier and one user, and a NET with its many suppliers and many users (NET:TSS = TSS:SAC). Just as a fundamental goal in time sharing has been to give each TSS user the illusion of having his own SAC,[1] a fundamental goal in networking is to give each user the impression that he is dealing with only one supplier no matter how many or whose services he uses.

Participants at the seminars also discussed a related objective, the desirability of making a NET "transparent." This goal is likewise a desideratum for time-sharing supervisors, most often observed in the breach but serving nevertheless as a gauge by which the success of a TSS can be measured. The success of a NET similarly might be measured by its transparency and apparent consistency across suppliers, along with its stability, reliability, efficiency, and ability to support its users and explain its services. The better a NET, the less the user is reminded that he is on it, the more it seems to him like the operation of one supplier, and the less he is aware that he is sharing its resources with other simultaneous users.

As the analogy of the geometric mean suggests, the complexity of the operation can go up geometrically as one moves from a SAC to a TSS to a NET. A NET presents special problems because of the involvement of many vendors, systems, and supplying organizations. Is it worth the trouble? Not everyone is convinced.[2] If everything offered by a NET could be provided as well by one TSS developed and operated under a single management roof, it would be difficult to understand the reason for wanting the NET. Of course, the same could be (and was) said of a SAC in relation to a TSS. But although some still grumble about the additional overhead or burden they believe time sharing imposes on their operation compared to batch processing, time sharing has become an accepted mode during the past decade without displacing batch processing as a mode of continuing importance. Time will tell if a similar verdict is to come down in the matter of networking versus time sharing.

Types of Networks

One speaker took a different tack in examining the meaning of "network." He considered the functions that networks perform (see Chapter 32), referring to the three most important functions as

- "Task-centered" operation, or the provision of services and facilities to accomplish specific jobs for users of the network.
- "Signal transport," or the physical movement of data from one place to another across the distances spanned by the network.
- "Communication facilitation," or the making possible of reliable, versatile, efficient communication between users and suppliers.

This functional breakdown suggests a somewhat parallel distinction among three different types of networks or network organizations that may or may not be separately distinguishable in network operations of the future:

- A *user-services* network.
- A *transmission* network.
- A *facilitating* network.

The user-services network can be taken to include the users of services, the suppliers of services, and the resources from which the services derive. The users and suppliers are joined not necessarily by physical links but by their mutual desire, commitment, and capability to share resources. To achieve a user-services network may require considerable change in attitudes, training, development of interpersonal contacts, and refinement of resources. The user-services network is fundamentally people-directed and task-oriented.

The transmission network, on the other hand, is primarily physical in nature. It consists of a set of communication facilities by which machines can pass data to one another. The facilities are generally automated message processors and high capacity cable, microwave, and satellite telecommunication channels, but in fact may also be ordinary dial-up telephone lines or even the U.S. mail. Recent changes in technology and regulatory attitudes are changing the picture for data transmission. The development of packet switching and other new transmission and switching technologies seems likely to bring down significantly the costs of high-capacity telecommunications, and a number of companies in the private sector may become common carriers to exploit the new technologies.

Mediating between the transmission and user-services networks are facilitating networks that may be likened to the broadcasting systems of radio and television. The functions they provide might include any or all of the following:

• Creating and enforcing standards, as for transmission codes.
• Establishing and implementing basic user protocols.
• Performing centralized accounting and billing.
• Furnishing documentation and general user support.
• Making a market for computing services.
• Supplying interface hardware and software, such as the IMPs, TIPs, Network Control Programs, and Host Interfaces of ARPANET.
• Providing communication services from the facilities of one or more transmission networks to the user-services networks, as communication services are provided by the broadcasting systems to the radio and television stations that are their affiliates.

The lines between these three kinds of network organizations cannot be cleanly drawn. For example, in the over-the-counter market system (NASDAQ), the Bunker-Ramo Corporation serves as a facilitating network, using standard facilities of the common carriers (transmission networks) to provide price quotations from dealers to brokers and dealers (user-services network). The National Association of Securities Dealers (NASD) performs an essential function in the provision of these services as a single organized body both representing and regulating the dealers and overseeing the services. Is its function one of a facilitating nature, a user-services nature, or both?

One way to attempt to resolve such quandaries is to divide the facilitating function into a manager part and an operator part, with the manager in control and responsible for funding, enforcing regulations, and the like, and the operator responsible for the operation and its satisfactory support. Thus Bunker-Ramo would be the operator and NASD the manager of NASDAQ. In the case of ARPANET, Bolt Beranek and Newman would be the operator and ARPA (currently) the manager. In the field of international satellites, COMSAT would be the operator and INTELSAT the manager. Which functional distinctions will best hold up for networking over time is difficult to know just now, but for purposes of this report we shall continue with the three-way classification into user-services, transmission, and facilitating networks.

Nature of Network Development
Although many participants saw the greatest immediate need as the organization of user-services networks, they recognized that the possibilities offered by the facilitating network could be key to other developments. The facilitating network has the capacity to reduce dramatically the "friction of space" in computing and to bring effective sharing within reach of a wide variety of geographically dispersed and organizationally dissimilar institutions. Once an effective facilitating network is established, it might offer the means to accommodate without major modification the addition or trial of a variety of user-services networks. If the same facilitating network accommodates both general computing services and science information services, for example, it could be useful to several different communities of interest.

Although participants felt that one, two, or at most a few facilitating networks might be entirely adequate for the needs of national networking, it seemed clear to them that a similarly small number of user-services networks would not suffice for all institutions of research and education. Higher education alone seems much too diverse in its requirements and much too heterogeneous in its political structure for all its institutions to get under the same tent. More likely is a variety of discipline-oriented and mission-oriented user-services networks. Some seminar participants referred to them as *subnets* and argued that institutions and their members should have the option of entering into or staying out of each subnet according to what it had to offer them compared to other alternatives. Discipline-oriented members might be expected to exercise considerable control of subnets in their discipline.

The concept of widespread pluralism in the development of user-services networks was not forbidding to participants so long as there was at least one large-scale facilitating network for them to associate with. Market forces should encourage discipline-oriented subnets to rely on a central facilitating network for their procedural and communication needs. The facilitating network would provide the glue to hold the sub-nets together.

Some participants were concerned that a trend toward discipline-oriented subnets could have a disruptive effect on the organization of computing in universities if discipline-oriented networks with economical telecommunication facilities tended to draw away computing sponsored by outside agencies. If this were the most efficient part of the university's computing load, as it might be, the university could find itself suffering doubly from a loss of sponsored revenue compounded by a comparatively more expensive and difficult residual computing load. The university could be the victim of a kind of "cream skimming," unless counter-measures were developed through careful planning and possibly restricting regulations.

A university might respond to a drain of computing dollars by eliminating its large-scale central computing system, as Harvard has done. The remaining load might be taken care of by less-expensive machines, including minicomputers. But if there were at least one stable, reliable set of general-purpose computing services available from a large-scale facilitating network, much of this load could be farmed out to external sources. In this case the university might be expected to convert its computing center into a user-services center that would contract through the facilitating network for information and computing services at the best price. This possibility suggested to a number of participants a whole-saler-retailer type of arrangement in the marketing and distribution of user services.

A Wholesaler-Retailer Marketplace
In the wholesaler-retailer concept, a number of large suppliers provide information and computing services to a greater number of smaller nodes on the network that serve as retail outlets for users in their districts. The University of Illinois operates such a retail outlet today based on a PDP/11-20, called a "minihost." The minihost provides a general-purpose port to the ARPANET, with printer, reader, punch, disc, tape, and graphic output device, in addition to providing line concentration for interactive terminals. The computer center of the University of California at San Diego provides what amounts to wholesale service to the Illinois group through the minihost. Its B6700 substitutes for a similar machine that the Illinois group used to operate itself. The Illinois group is said to make extensive use of several such wholesaler's computers on the ARPANET, which it views as a marketplace for user

services. Under the special pricing structure made available by ARPA, purchase cost of the Illinois minihost and network hardware are reported to be less than half of the bill for one year's computing; communication charges, less than 2 percent of the sum paid to wholesalers; and the total costs for computing services, less now than when Illinois operated its own B6700.

In a wholesale-retail system, the retailer provides local aid and information to customers and charges a markup to cover the costs of this support. One participant pointed out that present-day computer centers operate as combined wholesale-retail outlets. When joined with a facilitating network, they can provide retail outlets to local users for distant network wholesalers and can at the same time serve as wholesalers to the network. Since this participant regarded the economics of the two functions as fundamentally different, however, he expected to see an evolution toward a more clear-cut distinction between wholesaling and retailing in the future. He felt that some high-volume users with little need for user support would no longer be willing to pay the overhead burden currently charged for support services at many computing centers and would deal directly with specialized wholesale facilities, while the much larger group of users would continue to require good support services as supplied by the retailers.

This same participant pointed out that one service included in the support would be analogous in some ways to that provided by an air-freight forwarding company. The retailer would help the user obtain access to the wholesale service most appropriate to his needs and charge a fee that might be separated into a local component (for example, a lines-printed charge) and a general component, calculated as a percentage of the wholesaler's fee according to a volume discount schedule.

Degree of Central Management and Regulation
Allowing for the possibility that wholesalers will include university, federal, and privately owned centers, many participants felt that some form of regulation would be necessary to protect users and their organizations. Some favored regulation by means of admission rules and operating standards administered by either a network governing board or a strong trade association of network participants, somewhat in the style of the stock exchanges. Items for regulatory attention might include the following:
• The relation between aggregate cost and aggregate revenue for whole-

salers, and whether full cost recovery should be the basis for setting
prices.
- Nondiscrimination clauses and length of service assurances for network
  users to prevent the kind of situation wherein a low-priced wholesaler
  creates havoc by supplying network service for a year or two, then
  withdraws that service because of increased load at home or to accommodate favored clientele.
- Requirements for sufficient advance notice in the introduction of new
  resources or the addition of a network traffic load so great as to impact
  existing services, so that appropriate steps can be taken to alleviate
  expected insufficiencies and perturbations.

In considering other measures that an institution might take to maintain control of the quantity and quality of its computing and information services when it relies on outside sources, two alternative management approaches were debated. First was the model of "participatory
management," such as a consortium governing the network operation of
a market that is centrally managed. Here the institution participates
directly in network management under an agreement that calls for an
appropriate balance in influence among members of the consortium and
ensures that each institution gets its fair allocation of system resources.

The alternative is the "open market." Assuming the existence of one
or more facilitating networks, an institution can seek to protect its interests by contracting with multiple suppliers and threatening to shift its
"business" to a competitive supplier if service becomes unsatisfactory.
Contracts can be short-term or long-term in character and can reflect
the bargaining power and volume of the buying institution. The protection offered by this approach depends on whether the facilitating
network has achieved critical mass in terms of the number of alternative
suppliers it makes available to users, on whether basic user protocols
and programming systems have achieved adequate standardization across
the network, and on whether data files can be transferred from machine
to machine at reasonable cost and convenience. If current difficulties in
switching from one system to another are not overcome, this possibility
will be of more academic than practical interest.

Advocates of the open-market approach argued that consortia and other
participatory management systems could and probably would still exist
to share resources and set standards at a more detailed level than that
enforced by the facilitating networks. They regarded the critical factor
in developing the open market as freedom of entry of server nodes onto

the facilitating network. This would mean, in their view, that institutions seeking particular services (for example, APL) could shop around for them, and if an urgent unfilled demand existed or could be foreseen, someone would come forward to develop the desired service to meet it. They felt that a large-scale participatory management consortium, on the other hand, would tend toward "lowest-common-denominator" allocation of developmental resources. But they agreed that in either case, the existence of a broader market through networking could only be an improvement over the present fragmented national computing system.

Style of Computing Operations in Higher Education

Colleges and universities have traditionally developed computing and information services primarily for their own internal purposes. They have on occasion sold off their excess computing capacity to other schools and even to industrial customers, especially in periods of financial stress, but this has normally been considered a temporary measure that would be put aside as soon as local demand was restored or new funds were secured. In other words, the interests of the outside customer have come second to those of the institution, and stability of service for the outside customer has been constantly in jeopardy.

Perhaps this attitude has begun to change with the advent of regional computing networks, but some participants doubted that any college or university would ever want to put itself in the permanent position of having to give assurances or be accountable to sister institutions for the quality and reliability of their computing and information services. The California universities with ARPANET users are undoubtedly pleased to be receiving the additional revenue this outside business brings. But if this business were to double or triple, so that additional systems and personnel would be necessary to accommodate it, would they be as happy? Would they accept the additional business? Would inside users feel their interests were being compromised by the outsiders?

Unless colleges and universities are willing and able to alter past attitudes and loosen past constraints, it would seem unlikely to some participants that they could be considered as serious contenders for roles as routine suppliers of services in the networks of the future. Perhaps their roles as suppliers will be of a specialty nature and commercial firms will be the routine suppliers. The roles of colleges and universities as *users* need not be similarly qualified.

There are numerous other problems relating to the style of computing operations in higher education that received attention at the seminars. One, for example, was the difficulty of controlling the computer use of students. The problem is serious enough when students are using the computer at the local center. What happens when they have access to computer facilities throughout the country via a national network? That prospect, chilling to some, would challenge any accounting scheme.

Funding

Not surprisingly, a big question in many minds at the seminars was where the money for future networking development was to come from and how it was to be distributed. Some participants felt that institutions of research and education, given traditional kinds of governmental assistance and active participation by commercial firms, could find the means to advance networking themselves. But others, reflecting on the very real problems posed by resistance to change compounded by hard-pressed budgets, were convinced that some source of new money was essential for real progress to be made, if only to help provide start-up aid and incentives for the reorganization and toolup necessary.

Many participants saw networking as critical to the future health of research and education, and also to the strength of the American economy and the nation's competitive position overseas. From that premise, it seemed to them that the needed start-up money should be forthcoming from states and the federal government as a sound investment in the future, and that support should also be provided by the private sector with leadership coming from the large foundations. With respect to the government, unfortunately, the current funding posture gives no cause for great confidence in that possible source, at least in the near term.

Assuming the eventual availability of start-up money, the question shifts to how this money should be used. Some saw merit in consolidating work and finances in one major development project. Others found greater advantage in distributing grants in smaller amounts among many separate but coordinated developers. Some favored direct funding of suppliers, especially in the early stages of development, to enable new and improved network resources to be created and made operational as quickly as possible. Others preferred giving the money to users to augment their buying power, stimulate the market, and help ensure that resources meet user needs.

One of the most difficult problems discussed was the possible desirability of new patterns of funding and the implications these patterns might have within existing institutions. Network users tend to be aligned by disciplinary and other groupings that are orthogonal to their institutional affiliations. By virtue of the type of use he makes of a network, a professor of chemistry is primarily a crystallographer or theoretical chemist, not a member of the faculty of University X or a resident of State Y. In view of this fact, how should a chemistry subnet or other user-services network be funded? Should the universities and states be involved? Facilitating networks also cross institutional and state lines. How should their development be financed? What is the appropriate role for present-day institutions of research and education in relation to future networking developments? User groups, subnets, and consortia were among the organizational forms considered, but there were no pat answers. These are politically oriented questions. They deserve thoughtful study. Their satisfactory resolution may be critical to the future success of national networking.

Government Policies and Practices
Possible alteration to patterns of funding touches on the more general issue of government policy with respect to computers and information services in research and education. Such policy, insofar as it exists or appears to exist, has been a sore point over the years with university officials. It came under critical scrutiny and outright attack in several seminar discussions.

One concern expressed was with the federal government's practice of making certain large computing facilities available to special classes of users at greatly reduced rates or entirely free of charge. Participants spoke of facilities subsidized by the Atomic Energy Commission and the National Institutes of Health that were operated under a policy of limited access and preferential rates. The feeling was that this practice, if continued in an increased networking environment, would present unfair competition to nonsubsidized resources on the network, just as it presents unfair competition today to university computing centers. Other participants were more concerned with the impediment such a practice puts in the way of greater cooperation and sharing among universities. These people favored lifting current restrictions on the use of subsidized facilities and making them generally available, possibly under review by an external board set up by the Government Accounting Office to consider the requests and needs of a greatly expanded group of qualified users.

One speaker speculated that the reason networking had not developed further or sooner in higher education was because university computing centers were self-interested entrepreneurial units with little incentive to cooperate or yield any of their autonomy. Insofar as this characterization is valid, and a number of participants believed it is, these participants would trace its genesis to past government policies and practices. They argue that real progress in greater sharing and joint endeavor will not be possible until the various government agencies that support, affect, or regulate computing in research and education get together to work out consistent policies and coordinate their actions and decisions.

Many participants avowed that a well-planned and coordinated government-wide program to stimulate and support the development of information services and computing in research and education was long overdue. They regarded the movement toward networking as an opportunity to bring such a program about. One participant called computer technology one of the country's most important resources. He asserted that this technology contributes significantly to the country's economic strength both in the management of our complex society and as an exportable commodity. To his way of thinking, computer technology can maintain and even improve its contribution through the use of communications to increase the sharing of information and computer resources. He argued that the implementation of such resource sharing on a national scale in both the near and distant future should be a top priority item for the attention and coordinated action of state and federal governments.

It is often mentioned that one of the significant benefits of networking is the expanded opportunity it presents for the law of comparative advantage to take effect among user and supplier institutions. A similar principle may apply in the international area. The United States has a comparative advantage with respect to other countries in computer technology. It also has a comparative advantage in communications technology, but this advantage is somewhat muted by the strong nationalistic patterns in the supply of communications equipment and services. The merger of the nation's superior communications technology with its computer technology through networking may be an opportunity to strengthen the computing and information-processing resources that it brings to world commerce. In accordance with the principles of international trade, this should work to the benefit of all countries. It goes beyond the main subject and purpose of the seminars but is not unrelated and is intriguing to consider.

References

1. John McCarthy, "Time-Sharing Computer Systems," in Martin Greenberger, ed., *Computers and the World of the Future,* MIT Press and Wiley, New York, 1962, pp. 231–236; paperback edition, MIT Press, Cambridge, Mass., 1964, pp. 231–236.

2. Remarks by John Kemeny at the EDUCOM Spring 1973 meeting at the Harvard Graduate School of Business Administration, April 6, 1973.

3 Conclusions and Recommendations

Among 152 people chosen for the variety of their interests and backgrounds as well as for the strength of their intellects, it would have indeed been miraculous (not to say a bit suspicious) to find complete unanimity on any single point. But there was little danger of that. There was a healthy diversity of opinion throughout the seminars on most matters and absolute concurrence on none. Yet although no votes were taken and the impressions of four faculty members are fallible at best, there did seem to be enough agreement on certain points to warrant their being separated out and specially identified. They are recorded here as the conclusions and recommendations of the seminars on the basis of the particular significance that the faculty perceived was accorded them during the seminar discussions and in the planning and review meetings. Many other interesting recommendations by individuals and groups of individuals are scattered through the talks and workshop reports of Parts II, III, and IV.

Conclusions

The most basic conclusion became apparent by midway through the first seminar. It took on such force and clarity by the second seminar that it became a main theme in the faculty's summary remarks on that occasion. It received still greater emphasis in the third seminar. The conclusion is:

Computer networking must be acknowledged as an important new mode for obtaining information and computation. It is a real alternative that needs to be given serious attention in current planning and decision making.

To some the point will seem obvious. Yet the fact is that many institutions are not taking account of networks when they confer on whether or how to replace their main computer. Others may find the conclusion extreme. They should have been at the seminars to witness the sense of importance and general interest in networking that were so manifest there. Some feeling for this spirit may come from reading Parts II, III, and IV.

Articulation of the possibilities of computer networks is not what is new. That goes back to the early 1960s and before. Nor is the implementation and operation of networks new. Many working networks have been in evidence for several years now, both commercially and in universities. What *is* new is the unmistakable recognition, bordering on a sense of the inevitable, that networks are now practical and here to stay. The visionary and promotional phases of computer networks are over. It is time to sit up, take notice, and do some hard-nosed comparative analysis.[1]

A second conclusion that may be to many as self-apparent as the first has to do with the factors that hinder the fuller development of networking:

The major problems to be overcome in applying networks to research and education are political, organizational, and economic in nature rather than technological.

This is not to say that the hardware and software problems of linking computers and information systems are completely solved, any more than the payroll data-processing problem was completely solved in a technical sense in 1956. (It only seemed that way.) But technical problems are not the big bottlenecks that political, organizational, and economic issues are at present. Government agencies must learn to be consistent and coordinate their actions in setting policy for computing. Research and educational institutions must find ways to organize themselves (not just their computers) to work together for greater resource sharing. They must be willing to trust external agents (who warrant or can earn their trust) for the provision of computing and information services. The financial means must be found to help these agents bring the quality and reliability of their services to a level that wins the satisfaction of the using institutions who may be asked to yield some autonomy and subordinate or recast their own operations in deference to the network. Developing faith in the supplier will take time. But once the consumer is convinced the food store will always have milk, he will no longer feel the need to have his own cow.

The coming of age of networks takes on special significance in light of the dissatisfaction that participants expressed repeatedly throughout the seminar discussions with the present computing situation. There was a strong feeling that the current mode of autonomous, self-sufficient operation in the provision of computing and information services is frequently wasteful, deficient, and unresponsive to users' needs. The waste was said to come about because of duplication, redundancy of effort from one installation to another, incompatibilities, other impediments to the sharing of resources, and underutilization of installed equipment. Yet, simultaneously, there was thought to be much unsatisfied pent-up demand for computer and information services, as well as widespread unhappiness with the general level of documentation, program support, and user assistance. Complaints were voiced about the relative lack of uniform standards and the paucity of information on what programs and data are available and how to get and use them.

The human tendency, when beset by problems such as these, is to seek a savior in the next new technology—networks in this case. But the seminar participants were not so ready to delude themselves. The general feeling is expressed in the following conclusion:

Networking does not in and of itself offer a solution to current deficiencies. What it does offer is a promising vehicle with which to bring about important changes in user practices, institutional procedures, and government policy that can lead to effective solutions.

Thus, more critical than *whether* networking is developed and applied is *how* it is developed and applied. For example, networking emphasizes the need for standards and good documentation. Unless effective mechanisms are developed and strong measures taken in networking to ensure that suitable standards and documentation are developed, present inadequacies could get worse, not better. Third-generation computers were introduced into computing centers in the mid-to-late sixties to increase operating efficiency. They had just the opposite effect initially because of the excessive overhead burden imposed by their very elaborate operating systems. In addition, third-generation computers required costly and agonizing reprogramming of much operational software. To many, this blessed technological advance was a curse in disguise. Networking is not immune from similar problems, even though one might argue that its ability to link different systems together and make reprogramming unnecessary protects the user against major disruptions.

Recommendations

Participants had different opinions on how close the country was to being able to make national networking operational on a major scale, given that this was a desirable objective. A spectrum of views ranging across the following four positions emerged from the seminars:

1. We need a great deal more research before we can know the best direction to take. Simulation models need to be built and small-scale experiments undertaken to acquire the necessary data and to perform the required analyses.

2. Basically we already have as much operating information as we need

relating to systems efficiency, although it is largely unanalyzed. We should concentrate our data collection efforts on existing networks, conducting in-depth case studies, drawing lessons from past experiences, and trying to gain a deeper understanding than we have of the effectiveness (as opposed to the efficiency) of current networks and how they aid the users in their work.

3. We are ready to go ahead now with the creation of national facilitating and user-services networks but need detailed market research before developing a concrete plan. We need to know more about how many potentially interested institutions there are, where they are, what they are like, and what kinds of computers, user requirements, and user and administrative attitudes they have.

4. All we really have to do to create a facilitating network is to formulate a specific business plan and proceed to implement it. This process could be performed by any appropriate group with the necessary drive, initiative, intelligence, and resources.

Many participants believed the actual situation was closer to that represented by the first two positions than to the last two, although a certain amount of impatience and some despair was voiced at the thought of more studies. The general feeling can be expressed in the form of the following recommendation:

Additional research and discussion are required before deciding whether and how to launch a major national networking effort, but this should not be taken as reason to delay other activities that could help to illuminate the prospects and problems of large-scale national networking.

Concurrent with the study and discussion, it was felt that efforts should be mounted to collect data, analyze existing networks, and perform highly targeted experiments designed to investigate important issues and areas of uncertainty. Organizational activities, market research, and business planning also need not and should not be deferred. Quite to the contrary, participants considered this to be the right time for top-level decision makers at different institutions to meet and begin thinking about how large-scale networks might look and work. Officials must question seriously and honestly how well computing and information services are currently supplied, how well users are served, the historical reasons for the present mode of operation, how networks and other fast-advancing technologies might improve the operation, and what changes in government policy and institutional attitudes would be required or helpful in bringing about this improvement.

With this in mind, it is recommended that

Institutions of research and education should commit themselves to a comprehensive reexamination of how they supply and receive computing and information services and they should think imaginatively about future possibilities for networking that could help correct the problems and deficiencies that exist today.

The planning could be the basis for the formulation of funded programs by government and private foundations designed to assist the institutions move to a more effective mode of obtaining computing and information services in the years ahead. It could also serve as a concrete talking point in beginning the dialogue with users and university administrators, and it would likely provide stimulation, ideas, and guidance for specific experiments to be conducted.

A Planning Council

The general feeling was that now is the time for institutions of research and education to commit themselves to reexamine, compare, analyze, plan, and begin to organize and coordinate their actions with respect to computing and information services. This leads to the most action-directed and most encompassing recommendation of the present report. It is recommended that

A planning and organizing council on computing and information services in research and education should be formed to provide educational and research institutions with an organizational locus for continuing the study of networking possibilities, identifying and discussing current problems, dealing with funding sources, handling internal relations, and negotiating with one another.

The council should be a working group rather than an associational organization of representatives. Its membership should include people from institutions of higher education, research centers, information service groups, libraries, and laboratories (including those of profit-making corporations), selected on the basis of their experience, expertise, and ability to contribute not only to the deliberations and mission of the council but also to the national networking arrangements to which the activities of the council might lead. Since there should be no a priori presumption that national networking is the correct course of action, it

would be desirable to have some members of the council who question
the desirability of networking.

It is not clear how the council might best be organized. One possibility
would be for one or more institutional-oriented and one or more discipline-oriented organizations to form the council under their auspices: for
example, the American Association for the Advancement of Science,
the American Council of Education, the Association of American Universities, EDUCOM, certain professional societies, or other national
organizations. The council might be set up for a five-year term to

- Commission studies, tests, and experiments exploring the problems
  and prospects for national networking.
- Draft a plan for the formation of subnets and appoint panels to
  identify resources.
- Stimulate planning and organizational activities that will provide
  orderly transition to a large-scale facilitating network operation, if one
  is deemed desirable.
- Recruit entrepreneurial activists knowledgeable about networking to
  go out, answer questions, determine interest, and enlist support for
  the work of the council.
- Inform institutions about plans and future possibilities to help them
  prepare, adjust, and avoid disruption and surprise.
- Recommend an organizational structure for follow-up activities in
  the period beyond the life of the council.

The council should have a strong, experienced staff. Research and
experimentation conducted under the aegis of the council generally
would be performed by teams of individuals drawn from different institutions and supported by the council's staff. The members of the council
might be organizations and institutions, each served by two officials:
Normally one would be the president, chancellor, provost, senior vice-
president, or director at the topmost level of organizational authority at
a point where the financial, operating and planning responsibilities of
the organization meet. The second would be someone in a position that
brings him in contact with a broad cross section of the users of information and computing services; for example, the director of research, the
coordinator of computing resources, or the director of computing activities. Most of the ongoing work of the council would fall to these sec-
ond-level people who would see to it that their senior associates were kept
fully informed of activities and progress.

With the proviso that the council be a working group, its organizational
membership should be as broadly based as possible. It should be sup-

ported by a substantial financial contribution from each member that expresses the member's commitment and gives the council the financial base it must have to commence its activities. The expression of commitment and state of solvency thus obtained would help the council attract the important additional funds from other sources needed to carry forward its activities on a scale commensurate with the magnitude of the endeavor. State legislatures, federal agencies, and private foundations are keenly aware today of the financial problems of educational institutions. They are looking for ways to improve educational efficiency. The work of the council is certain to be of special interest to them.

As important as the formation and work of the council may be, it must not be set up or operated in a way that blocks or discourages separate or prior organizational efforts. Neighboring institutions that wish to enter into an agreement to work together and share facilities before the council is formed should not be put off or dissuaded because of the imminence of the council's establishment. To be successful, the council must act as a stimulant and guiding force, not a depressant or inhibitor.

Organizing User-Services Networks

With respect to organizing user-services networks, there was strong agreement. It is recommended that

Organizational efforts to locate and bring together disciplinary, mission-oriented, and other logical groupings of similarly interested or motivated individuals whom networking can potentially benefit should get under way without delay.

These groupings, termed potential user groups (PUGs) by some participants, are already formed or forming in certain fields. Together with the agreements and mutual self-interests that unite them, the PUGs provide a basic element of the subnets or user-services networks. The other major elements are the suppliers and their resources. It is further recommended that

Suppliers and resources potentially useful for networking should be identified, and where the desired resources do not now exist, effective incentives and development funds should be provided to stimulate their production or adaptation to networking.

The use of advisory committees and of discipline and mission panels selected from the PUGs would help in identifying these resources. In some cases, the panels might suggest a known, respected resource such as the BIOMED package of statistical routines at UCLA; in other cases, they might propose an innovative potential supplier such as Project MAC at MIT. Often a user group such as the CACHE committee of chemical engineers will indicate what resources it needs or desires and where they might best be located or developed. Then protocols can be agreed on, the type of service desired can be determined, specifications for the necessary hardware and software can be developed, appropriate financing can be arranged, and negotiations with prospective suppliers or producers, if necessary, can be undertaken. This form of planning requires a coordinating body to take a leadership role in encouraging users to come together and in seeing to it that protocols, standards, review procedures, and negotiations are internally consistent and likely to conform to the requirements of a facilitating network, once established. The Planning Council would be in a good position to fill this role.

The feeling was that the role of the council in organizing user-services networks should be helpful but not central. That is, the council might draft a plan for organizing subnets and make it generally available and visible, but it should not set itself up as an operating agency for subnet formation. Participants felt it would be important to maintain an open and nonexclusive spirit in the organization and operation of the subnets. It therefore is recommended that

A modular structure and noncompulsory, nonexclusive membership policy should be adopted to allow institutions and their faculty members to consider the advantages and disadvantages of joining each subnet individually, and independently of their joining or not joinng the other subnets.

The strategy of adding subnets one at a time and permitting institutions and their members to pick and choose according to interests, benefits, and costs will give users maximum flexibility at the same time that it affords system designers a graduated schedule of development that lends itself to continued checking and evaluation.

Stimulating the Creation of Facilitating Networks

Participants fully recognized the importance of an effective facilitating network in the formation of subnets, even though they did not want to wait in beginning to form subnets until a national facilitating network was in operation. They understood that the successful development of one or more national facilitating networks, along with the transmission networks they use for digital communications, would be crucial in spawning widespread networking activity. They were convinced that to keep its customers happy and attract new ones the facilitating network would have to make available a high grade of user support and a strong human organization that keeps the customer's best interests in mind, stays honest and fair in its charges, is responsive to all on an equitable basis, and works diligently to achieve reliability, availability, and understanding in its services and support functions.

The Planning Council will want to give some hard imaginative thought to how the creation of facilitating networks might be stimulated and aided. This includes the development of incentives and other mechanisms to encourage commercial firms and the producers and suppliers of services to work with the facilitating networks to create and maintain a high-level of documentation and other user support. One such mechanism would be a page-charge type of subsidy included in government-funded development contracts, providing for documentation and maintenance services well into the future. Some participants believed that documentation should eventually be self-supporting, but special incentives would help initially attract commercial firms into the operation of the facilitating network. The experience and talents of these firms in the marketing and distribution of services would be extremely helpful. Participants agreed that it is important that the commercial sector be invited into national networking developments in a way that enlists its cooperation and makes full use of its special abilities.

Although parallels do exist, facilitating networks are basically a new form of organization and, as such, deserve careful and innovative study. Much of the research on networking conducted under the aegis of the Planning Council and elsewhere should serve to provide a necessary understanding and firm foundation for the creation of facilitating networks.

Types of Research Needed
The kinds of research that would help identify and illuminate networking issues fall into three categories:
- Market research, economic analyses, technical investigations, and simulation studies to examine a variety of economic and technical questions.
- Trial operational efforts to test the economic viability and practical validity of the networking concept and to gather data.
- Controlled experiments in the field to provide planning information,

explore uncertain areas, and accumulate experience on the effects and usefulness of networking.

With respect to the first category, studies are needed to analyze the economics of networking and to advance the state of the art, especially with regard to the refinement of user protocols, software interfaces, hardware/software trade-offs, program control, and systemwide file management. Simulation and gaming studies are also needed to explore the workability of user-services networks and to assess their political, organizational, and economic implications, especially in higher education. These studies could also be used to deepen our understanding of what is involved in forming and running a facilitating network.

With respect to the second category of research, if the autonomous, self-sufficient mode of supplying computing and information services is relatively inefficient compared to what networking has to offer, if there is indeed an enormous amount of computer overcapacity in higher education, and if at the same time there is a sizable pent-up demand for computer services—if these conclusions are correct—then it should be possible to design some reasonably straightforward trial operations that demonstrate the savings and market equilibration possible from greater networking. These trial operations should not be excessively difficult to arrange in that all parties involved supposedly would have something to gain. The tests need not be tightly controlled and should require only modest funding at the start to aid the parties over the transition.

A number of such trial operations have already been tried, especially by the regional networks, with mixed results. More trials are needed. In short, the most direct type of investigation to perform is the trial operation designed to test the proposition that networking is more satisfactory or economical for delivering certain classes of computer and information services than is autonomous, self-sufficient operation. Parties to the test would be expected to modify their modes of operation on an assumed long-term basis. The test requires a major commitment by these parties and a willingness to accept the risk and considerable inconvenience of a negative outcome. Additional funding may be required later in the operation to ease disruptions and maintain stability.

One such trial operation might be to link a number of large underutilized computers together and attempt to use them to provide computing services to a group of institutions that either do not now have their own computing centers or who like Harvard have closed down their computing centers to buy services outside. A trial operation of this nature is currently being attempted on ARPANET. The services offered by the system of computers would have to be established and maintained on an effectively permanent basis so far as the users were concerned and could be made gradually more sophisticated as the needs of the users developed. The buying institutions would pay for the services from their own funds to provide a valid test of whether this arrangement was indeed superior economically as well as in terms of services received to the alternative of each of the institutions developing or maintaining its own independent computer capability.

A second trial operation would be to interconnect a number of regional networks into a trial national network that would make the services of a regional network available to users in other regions. It would be interesting to see whether the addition of supply sources stimulated a separation of function and specialization among the networks or whether each continued operating pretty much as before the interconnection but with a larger clientele and possibly a more competitive orientation. Some believe that anything accomplished by interconnection of two regional networks would be better and more efficiently achieved by expansion of one or the other. The proposition needs to be tested.

The results of such tests, depending on their outcomes, would be useful either in helping to make a case to state legislatures and university administrations regarding the cost effectiveness of present operations, or in indicating that the advantages of national networking, such as they are, lie along dimensions other than those of greater economy and efficiency. To be convincing, the tests should be supervised and the results evaluated by an objective third party.

The third kind of research needed consists of controlled experiments designed to explore the nature of the networking mode, the problems it presents, its effect on the user, the new and expanded services it makes possible, their usefulness, and their potential impact on research and education. This type of investigation may not be as natural to arrange as the trial operation and may have to be substantially funded to provide incentives for those involved to depart from their customary practices without the prospect of immediate benefit. But the risk assumed is less than it is with the trial operation, and the measures required are more easily reversible.

One such experiment might be exploring whether users who are accustomed to having their own centers close at hand can be made at least as comfortable and content with a remote network as they are with their local systems. The attempt would be to make the network as responsive and transparent to the user as possible. The experiment might be conducted by up to twenty institutions, each experienced and well endowed in computing, each with something to gain from a network and with something to offer. In classical economic terms, the network would be used to foster a division of labor among the institutions according to the principle of comparative advantage. Users and suppliers would be encouraged to trade freely and to gain full benefit from the specializations and improved markets the network and division of labor made possible. To be interesting as a network experiment that provides more than a convenient mechanism for switching and communication, most users should have routine need for the services of more than one and preferably several suppliers.

A related experiment would be to investigate how effectively individual information-processing jobs could be functionally divided and allocated among two or more specialty centers in a network. An example is the social scientist performing a statistical study of sampled or selected data from several large data files: one of cross-sectional census data, another of individual consumer behavior, a third of personal public-opinion surveys, and a fourth of aggregate time series of national economic indices. Assume, as is actually the case, that a different center holds and maintains each of the data files. Suppose the programs of each data center permit the social scientist to select or sample data of relevance to his study but not to analyze it extensively or conveniently. Suppose also that a fifth center, geographically removed from the four data centers, provides a highly flexible, easy-to-use programming system for general analysis. The job of the network would be to furnish the general user, normally at still another (sixth) location, with a simple means for

- Causing each of the five data centers to extract the desired data.
- Having the selected data transmitted to the analysis center.
- Causing this center to perform the analysis according to the user's specifications as spelled out from an on-line terminal.

The objective of the experiment would be to see whether such an arrangement could be implemented in a form convenient and satisfactory to a wide class of users and whether it would be worth the trouble.

Still another example of an exploratory experiment is a version of an experiment that has been done many times over in regional networks: a trial run of 1,000 engineering undergraduates at two or more schools through a beginning programming course using another school's computer. This experiment can investigate whether it is possible and efficient to provide the type of remote coordination and personal user services important in extensive networking operations over great distances. That is, it can test the proposition that "remote" need not mean "isolated" in networking any more than in time sharing. So far in these experiments, the problem has been the heavy cost required for adequate support. But there is room for new ideas and more imaginative use of the technology. If successful, such a project could serve as a useful demonstration model.

Of the types of research needed, the controlled experiments would be the most complex, politically sensitive, and heavily financed, but would tend to be reversible in nature, whereas the trial operations would have modest external funding requirements, at least initially, and would be undertaken with an assumed permanency. The studies of existing networks, economic analyses, technical investigations, and simulations could be supported in the traditional way.

Reference

1. Martin Greenberger, "A Time for Commitment," *EDUCOM*, Vol. 8, No. 1, Spring 1973, pp. 2–7.

II User Characteristics and Needs

Part II contains the papers and workshop discussions of the first seminar on user characteristics and needs, their common elements, their significant differences, and their meaning for networking. Participants ask

• What are the special requirements of each class of user? What do users need that they do not now have? Where are the gaps? What new resources and services would help users do their work better? Where is the expertise? How can it best be mobilized? Where is it closest to being mobilized? What are the opportunities and incentives for sharing and collaboration? How can networking help?

• What should be the primary goals in the design of computer networks, in the forming of organizations, and in the origination of projects and experiments? What are the central policy issues and administrative problems in developing networks for sharing resources among users from the same and different fields? How should NSF's expanded research program be directed? What kinds of projects should be encouraged?

There are seven background papers. Don Aufenkamp discusses the NSF network initiative to explore possibilities for a national science computer network (Chapter 4). J. C. R. Licklider reflects on the potential contribution of computer networking to research and education (Chapter 5). Robert Kahn presents the status and plans of the ARPANET (Chapter 6). Max Beeré takes up commercial time-sharing networks using available common carrier facilities (Chapter 7). Norman Nielsen presents some lessons learned from a comprehensive study of regional networks (Chapter 8). Robert Ashenhurst examines the place for hierarchical computing in national networks (Chapter 9). Dennis Fife treats the principal issues being studied by the National Bureau of Standards in its work under the NSF program (Chapter 10).

Four additional papers serve as position statements to help start the four workshop discussions. Walter Hamilton examines use of the computer for large-scale algorithmic computation and numerical analysis as in the basic sciences (Chapter 11). Wilfred Dixon takes up use of the computer for the construction of large numerical databanks, statistical analysis, and behavioral modeling as in the social and biomedical sciences (Chapter 13). Herbert Teager considers use of the computer for interactive computing, on-line laboratory control, and other applications in medicine and medical research (Chapter 15). Robert Hayes treats use of the computer for the inventorying of knowledge, bibliographic maintenance, and information classification and retrieval as in library work (Chapter 17).

Workshop 1 is concerned with how computer network technology should be used to advance research in the chemical, physical, and biological sciences (Chapter 12). The starting point is traditional computer applications such as the processing of experimental data, numerical methods for solution of differential equations, large-scale model formulation, algorithmic solution, stochastic representation, and man-machine dialogue using active graphics. The workshop examines research opportunities made possible by the design and creation of new computer facilities.

In contrast to many physical scientists, the behavioral scientist does not typically design his own tools or methods in preparing a problem for a computer. He relies heavily on techniques and program packages developed by others. He is more a shopper than a manufacturer when it comes to software and program services. He often works with and has to maintain large numerical data files, performing extensive statistical analyses on these data and incorporating his theoretical results in behavioral models that he runs as simulations. In all these activities, he needs the support and assistance of others. Workshop 2 sets out to explore this kind of use of the computer with an eye to determining how networking might be able to advance and facilitate it (Chapter 14).

Workshop 3 deals with the special needs and opportunities for a network of users with real-time requirements (Chapter 16). Real-time demands range from on-line biomedical laboratories to on-line registration of students. Examples include an experimentalist tracking laboratory data, a teacher using computer-based instruction to support regularly scheduled class activity, a librarian operating an on-line information retrieval system, and administrators relying upon a responsive system to meet specific deadlines. Prime criteria are simple and reliable functioning and guaranteed availability on schedule. Users are often not particularly knowledgeable or sympathetic to computer functions, and the operation may require continuous maintenance due to the changing logic of the tasks. But there are opportunities to take advantage of the software developments of others, consolidate maintenance effort, provide a reliable backup system, and allow testing and development at one system while giving service at another. Possible problems stem from the propensity of users to insist on local control, the need for prompt documentation, and the idiosyncrasies of particular parties. Workshop 3 tries to identify useful, joint efforts that can lead to an improved understanding of the payoffs and problems of network cooperation for real-time users and current gaps that a network might be able to fill.

Workshop 4 is concerned with systems designed for the storage and selective retrieval of information from very large textual data files (Chapter 18). The emphasis is on systems that incorporate text-oriented processing for indexing and searching operations, provide information (like *Chemical Abstracts*), and allow users to develop their own file structures and retrieval protocols. The workshop attempts to identify a range of applications to determine the probable characteristics of the user populations for each and assess the economic and/or technological possibilities offered through networking.

4

D. DON AUFENKAMP
National Science Foundation

NSF Activities in Networking for Science

The National Science Foundation is embarking on a complex and far-reaching attempt to develop the resource-sharing potential offered by the coupling of computers and communications. If the concept of a national science computer network proves to be viable, it will not be merely because the Foundation announced a program. It will be the result of the cooperation of many people in response to genuine needs of the scientific and educational community.

The series of working seminars conducted by EDUCOM is a key element in this undertaking. It is designed to bring out the underlying issues and considerations that must be resolved in developing the potential of national networks in the support of research and education.

It may be helpful to review some of the NSF's activities related to a national science computer network and to provide a perspective on the program under which these activities, including this seminar, are being conducted. The overall objective of this program is to provide specific information concerning the feasibility of a national network that would provide its users with access to computing facilities, science information systems specialized data banks, and other computer-based resources regardless of location. Such a network would have profound implications. In particular, we envision a much closer integration of science information services with the usual academic computing than has been realized to date. In fact, a much closer coupling of computing and information systems would open new worlds for both research and its timely dissemination. In recognition of such developments, the Foundation's Office of Computing Activities and Office of Science Information Service are cooperating in mounting this program. But clearly the initiative reaches into many other parts of the Foundation and into other agencies and groups as well.

For many years the Foundation has sponsored research, development, and special studies in connection with resource sharing, user services, and network technology. This network initiative is building on these efforts. Under the Institutional Computing Services program (ICS), whose lifetime spanned the 1960s, the Foundation supported not only the development of individual campus computing facilities but also facilities to be shared among institutions. The Triangle Universities Computation

Center in North Carolina is one example see (Chapter 24). Of more recent vintage is the Regional Cooperative Computing Activities Program, which is providing additional understanding of some of the problems of developing computer facilities for shared use. Approximately 300 institutions of higher education, that is over 10 percent of the total in the United States, are involved in this program under some thirty regional computing networks. Norman Nielsen describes this effort (see Chapter 8; also Chapters 30 and 34). The CONDUIT project is a "soft" network involving five of the regional centers to examine the feasibility of transporting computer-based materials for use in higher education. Ronald Blum is director of that project.

One of the arguments for a national science computer network is that specialized facilities could be available for given disciplines or classes of computations. It is, in fact, much easier to talk about special resources for networks than the much more nebulous group of users of such resources. The National Center for Atmospheric Research, with its large-scale computing facility, is already well known. The Computer Research Center for Economics and Management Science of the National Bureau of Economic Research was established more recently with NSF support. This center conducts research in computer-based methodology for applications in the disciplines cited. Another resource with national import is The Clearinghouse and Laboratory for Census Data, established at DUALabs under contract with the Center for Research Libraries.

The Foundation, through its Office of Science Information Service, has also been supporting system development through the scientific professional societies and through the university sector. The society-based systems produce the information banks; the university-centered systems provide the distribution outlets to the user community. The Chemical Abstracts Service (CAS) of the American Chemical Society is one such computer-based resource. The system under development at the University of Georgia is an example of a broadly based system, with a primary focus in chemistry and the biological sciences. Lehigh University is cooperating with Georgia in still another project. I should also mention the cooperative program involving the Environmental Protection Agency and the science information centers at Lehigh, Georgia, and the Ohio State University.

Next is the area of computer communications. The research and development carried on with the support of the Advanced Research Projects Agency and by the commercial sector is well known (see Chapters 6 and 7). Foundation support for research is through programs in Engineering and Computer Science. The MERIT network in Michigan is an example of an NSF project involving a close linkage of three major computer centers. There is no need to belabor this topic of computer communications. The important point is that recent advances in technology are adding a new dimension to efforts to improve the sharing of computer-based resources. Computer networks, both commercial and government supported, that are nationwide in scope are in being. The technology has advanced to the point where the feasibility of a national network of computing resources that is transparent to the user is no longer an issue. Today it might be implemented with land lines, tomorrow with satellite-based communications.

But a great many problems do remain to be addressed of how institutions, and individual researchers and instructors as well, can avail themselves of network resources in the current complex computing and science information environment. These problems are not primarily technical in origin, although clearly the technology plays an important role. Rather they are organizational, political, and economic.

A network in this sense, then, is much more a question of people and resources than it is one of sophisticated computer communications. The Networking for Science Program of the Foundation is emphasizing these organizational, political, and economic questions, and it is here, particularly, where we call upon your help. It is important, then, to stress that this initiative is complementary, to the technology. The Foundation is not proposing to implement and operate a major computer communications network. But the support of resources for networks is clearly a continuing NSF function, and we do suggest, too, that an appropriate trial network should be an integral part of the research program.

These issues and considerations related to the development of a national science computer network include user characteristics and needs, organizational matters, and operations and financing. One of the considerations that arises is network management. Its function in achieving a viable and effective service cannot be overemphasized. Network participants, whether they are providers or users of resources, must agree on conventions and protocols. What obligations should the network have toward institutions and individual users, and conversely? What organizational structures would be appropriate? What substructures would be desirable? What kinds of policy and advisory bodies would be needed to cope with problems of sharing risks and liabilities? What bodies would be needed in connection with the evolution of the network? Most of the discussion seems to center on how to join a network. But once the commitments, perhaps heavy, to provide network services are made, what would happen if an institution chose to withdraw or could no longer make its resources available—particularly if users elsewhere had become dependent on these facilities for special systems or data? All these questions should be addressed carefully. The National Bureau of Standards, under Foundation support, is identifying tasks associated with setting up and operating a national resource-sharing network and is evaluating alternative strategies for implementing and operating it under various managerial structures. See Dennis Fife's contribution (Chapter 10) on these efforts.

The need for special resources and services has already been mentioned as one argument for networks. But what are the user characteristics and needs for research computing, for science computing services, and for educational computing? What special coupling of these functions can be brought about? One study currently under way through the National Academy of Sciences, with Foundation support, concerns the feasibility of a national laboratory for theoretical chemistry. Some specialists believe, for example, that appropriate use of large-scale computers can now bring about advances in quantum chemistry easier than can the traditional laboratory approach. A report is forthcoming that addresses the questions of such a national computational laboratory as well as the broader questions of interfacing it with the scientific community.[1] Harrison Shull has prepared some selections from these discussions (see Chapter 28).

Another study has been conducted at the University of Kansas under the leadership of Drs. Walter and Sally Sedelow to develop the concept of a national center or network for the broadly based interdisciplinary area of computational research on language. A report has been issued on this study.[2]

In still another project, this one addressing needs for user services, Frank Harris at the University of Utah is designing software to facilitate the use of theoretical calculations by workers in all fields of chemistry, including those not expert either in details of the theoretical analysis or in digital computation. The project provides for consultations regarding distributed software and pilot use of remote computing facilities directed at an eventual public use of computer networks. I will also mention hierarchical computing, in which a researcher has access to a number of levels. (Robert Ashenhurst discusses this subject in Chapter 9.)

The entire area of software testing and distribution is one for which a national network would have special meaning. As has been stated often, such a network would facilitate the transfer of software from machine to machine but would also reduce the need for doing so. Certain centers could assume responsibility for specialized languages and applications packages. The National Activities to Test Software (NATS) project is one example of such a specialized service. It is a prototype effort involving Argonne National Laboratory, the University of Texas, and Stanford University to evaluate, certify, and disseminate mathematical software. Sixteen test sites are associated with the project, including one in Canada and the Nottingham Algorithm Group in England.

This discussion brings us to the consideration of special-interest groups for more effective communication and sharing of network resources. This is clearly one of the "people" problems of networks. To emphasize this aspect, think of the network as a list of computer telephone numbers, or perhaps only one number with a great deal of expertise behind it. A key element in effective communication for network resource sharing would be extensions of the Network Information Center concept for the ARPANET developed at Stanford Research Institute.

Economics and financing of networks are also of special concern, but space precludes adequate discussion. In the long term a network of computer resources and services should be self-sustaining in that users of the network should bear the costs. We have one long-term project under way on the economics of computer communication networks under Donald Dunn at Stanford University. This project is working within the framework of the three principal dimensions of computer networks: classes of communication technologies, classes of services, and levels of analysis (market, demand, and so on).

The impact of a major national computer network on campus computing centers is a particular concern to the National Science Foundation, since it supports institutions as well as individual researchers. Considerations of campus computing are complex and often extend beyond basic economic concerns. Careful planning would be necessary before a major network could be superimposed on these centers.

According to a survey by the Southern Regional Education Board, expenditures for computing were $472, 000, 000 in fiscal year 1970 and $550,000,000 in fiscal year 1971. One of the arguments for a national network is that it would offer an economically attractive alternative to the usual practice of maintaining largely self-sufficient campus centers. What percentage savings in computing expenditures would prompt an institution to reduce its campus facilities in favor of network services? Would 10 percent be sufficient? Probably not. Would 20 percent suffice?

Possibly. But how would a decision be reached? And what would be an appropriate course of action in the event the network services were judged unsatisfactory? One project at the University of Denver is studying alternative approaches to the management and financing of academic computing centers. The sharing of facilities by means of a network is one alternative. Again, a report is forthcoming, and it should assist college and university administrations in dealing with the complex and growing problems of the role, management, and financing of campus computing facilities. One of the trends we see is an increasing activity on the part of state agencies to take direct control of computers, including those at public academic institutions.

Clearly, there are many dimensions and facets to the conceptual development of a national science computer network. Such a network would probably raise as many problems as it would solve. The key to this development is the subject of user characteristics and needs. I have provided at best only a brief overview. We look to the efforts of others to provide the structure and substance.

References

1. *A Study of a National Laboratory for Computation in Chemistry,* National Academy of Sciences, National Research Council, Washington, D.C., in press.

2. Sally Yeates Sedelow and Walter A. Sedelow, Jr., *Language Research and the Computer,* National Technical Information Service (NTIS), Department of Commerce, Springfield, Va., 1973.

5

J. C. R. LICKLIDER
Massachusetts Institute of
Technology

Potential of Networking
for Research and Education

Let me begin with this statement: "The ultimate purpose of a computer-communications network in our times is to provide a creative environment for people to interact in. By 'creative,' I mean a network which has great diversity and thus allows for freedom of choice and which generates a maximum of interaction between people and their intellectual surroundings."

This statement is modified from one by Lawrence Halprin.[1] Where my version says "computer-communications network," Halprin's original said "city." Where I say "interact," it said "live." And where I say "intellectual," it said "urban." Except for those small differences, the idea is the same. People will literally live in computer-communication networks of the future. They will work and play in computer-communication networks. Perhaps they will do so, to borrow another phrase from Halprin, "without much awareness of which is which: work or play." Most people's interactions with other people and with their intellectual surroundings will be mediated by networks. I think we all appreciate that that is true. We may disagree a bit about the time scale but not about the eventuality.

I think that we are five or six years into a sociotechnological revolution that will continue for ten or fifteen more years before we can figure out the general outline of the ensuing regime. The revolution will change the basic paradigm of human communication and, of course, computer-computer communication. It will change the paradigm both technically —as with the development of packet communications—and socially. Person-to-person communication will develop in such a way that it is routinely supported by access to data bases and models. Indeed, thinking will develop in somewhat the same way, being supported much more intimately than it now is by data bases and models. Interpersonal communication, which is essentially a thinking together, will derive much benefit from its mediation by an active network that will support it with processing and with memory.

The network revolution will also greatly change the way we use computers. I put the use of computers second, rather than first, because I think the impact of the network on human communication will be the more far-reaching of the two impacts if they are, indeed, separable. But the impact on how we use computers will be very great. My friend Edward Fredkin used to say around 1960 that one day computation, like electrical power, would come out of a wall socket and that users of it would not be much concerned about exactly where or from what kind of computer it came. I think that the network will simply put processing and storage at our disposal wherever we are and in whatever kinds of activities we are engaged.

It is worthwhile considering what fraction of the American population will interact with computers. I think the fraction fifteen years from now will be between 10 and 25 percent. I do not mean, of course, that that large a fraction will be programmers or will have jobs as closely tied to computers as are, for example, those of airline ticket agents. But many people will work directly with computers in the office, and many will use computers, probably mainly through telecommunication channels, in the home.

In my view, the technological feasibility of computer-communication networks is essentially established.

The problem of economic feasibility seems to rest on only one assumption—that a large computer-communication network, such as the ARPA network, can be brought into a state of use in which at least one-third of its capacity to transmit bits is filled on the average twenty-four-hour day.

That leaves the factor of social feasibility, and here I refer to all the questions of administration, management, and change of procedure and practice in individual and in organizational activities to which Don Aufenkamp refers in Chapter 4. There is the main factor of uncertainty. The social factor is indeed a source of concern to me and to everyone else who is interested in seeing networks have a chance to do some good for the world. If we consider the hodgepodge of undistinguished buildings in our cities, or the problems of environmental pollution that we have all been considering these last few years, or the vicissitudes of international relations, or any of the other major problems that have to be solved by by people working together in a cooperative way, we may not feel strongly reassured that the social factor will not grossly delay or distort the course of computer-communication networks.

If the social factor permits its development, what will the computer-communication network be?

First, it will be a "network of networks." It will not be a single, fixed system of interconnections among a set of computers and terminals. Rather it will be a continually changing configuration in which many subordinate networks can be discerned: some government, some private, some proprietary, some open, and so on. The connections will be ephemeral, changing with time, adapting—within an extensive and pervasive system of electronic roadways.

Second, it will, in due course if not at the outset, be a virtual network in the sense that one will not have to think much about the geographical distribution of its parts. In a good library system, if the requested document is not in the local holdings, it is located, borrowed, and delivered to the subscriber just as though (except for the factors of delay and expense) it were on the local shelves. In the computer network, the incremental delay associated with going far afield for information or for processing may not even be noticeable and the incremental cost may be quite tolerable. In any event, for many purposes it seems essential that the network present a uniform array of protocols and procedures to its users.

The computer-communications network of the future, then, will be a virtual network of virtual networks that connects people and their desklike consoles to other people and their desklike consoles and the informational systems and services. We should consider briefly what some of the informational systems and services will be.

By emphasizing person-to-person communication, I have implied that the services will include a kind of computer-supported telephone service and a mail service. The fact that processing and storage are available in the network will make it possible to introduce variants of telephone and mail that offer the best features of both. Computer-supported teleconferences and geographically distributed seminars hold great potential that we should try to develop. For another pattern of communication activity that I think I will develop, I have a name to propose. It is the "missing." A missing is like a meeting except that the participants do not meet or even interact at the same time. The focus of a missing is a data base. At the beginning of the missing the data base contains only background material. At the end, the data base contains the product of the interaction of the participants. The essential activity of the missing is carried on by the participants individually. Each logs into the data base from time to time, perhaps once a day, over a month, examines the new contributions on the topics in which he is interested, inserts his own comments into other participants' contributions, and adds a note or two or a major paper as the spirit moves him. He examines the summaries maintained by the secretaries of the missing and looks over the agenda of official business, voting on anything that requires his vote. He may address communications to other individual participants or to whatever subsets of the membership he likes. But he will do all this at times largely at his own choosing and without having to travel.

Other services and systems made available by the network include specialized assistance in various fields. Let me use MACSYMA to exemplify the concept of specialized assistance, emphasizing, in doing so, that mathematics is one of the few fields in which the concept has yet been implemented enough so that one can say the service actually exists. MACSYMA is a repository for knowledge about mathematics. At present it is very good at handling symbolic expressions of most of the kinds encountered in analysis. It is very good, for example, at symbolic integration. MACSYMA is proving very helpful to those who have been testing it as a practical tool in substantive work. I think that, even in five or ten years, many appliers of mathematics all over the country will be doing their work on-line through the network to a descendant of MACSYMA.

DENDRAL, as most of you know, is a program that embodies in the field of mass spectrometry a considerable amount of artificial intelligence. It is very sophisticated in interpreting the data provided by mass spectrometers and certain auxiliary data. Taken together, MACSYMA and DENDRAL constitute a large fraction of the total accomplishment to date in embodying specialized technical competence into computer programs. There are hundreds of fields in which such a program would be useful. Indeed, as my colleague Robert Fano has pointed out, the main hope for providing complex services of high quality to a large number of users is to incorporate those services into programs.

In contrast with the specialized assistance services, which are essentially organizations of procedure and data, are services that consist essentially of data. The network will make available data bases, sets, and elements of great variety. From the academic point of view, the role of data is crucial in behavioral and social science research, in the field of medical records, and in history and law. The task of making data not only available but worthwhile is formidable. As we all know, there is much more to it than punching holes in cards. Before data will be useful in the network they will have to be carefully checked. Computer-processible code books and directories will also have to be prepared, and arrangements for consultation and maintenance will have to be set up. The realities of data are understood in a few fields, such as political science, but those realities will be magnified manyfold by the rise of computer-communication networks.

Eventually the entire body of knowledge that now resides passively in libraries will become an active organism within the network, a continually self-scrutinizing and self-organizing complex of procedures and data sets. The estimated information measure of present library holdings, 10^{15} bits, provides a rough indication of the magnitude of network data systems of the distant future. At present, network technology seems about ready to tackle a 10^{11}-bits data base (in meteorology) with the aid of an on-line trillion-bit store. The Social Security data base, and about 10^{11} bits on magnetic tape, and the ten largest airline reservation systems, each on the order of 10^{10} bits on-line on discs, provide other gauges of the present state of the art.

In the latter part of this decade and in the next, network-accessible data bases will arise in many areas of government, business, and private life. Reservation services will include most hotels, motels, restaurants, theaters, and sports events, and advisory services will be available on-line along with reservations and ticketing. On-line shopping, linked to cable television and to all manner of consulting services and discount clubs, will create something of an upheaval in retailing. Urban and regional planners will have access to data on the number of vehicles of various types on various highways as a function of time of day and time of year—and so on. The network will deal with a great and diverse array of facts and figures, but they will have to be acquired, organized, and made processible before the network will have the content that is just as vital as its capability to store, process, and transmit.

In speaking of MACSYMA and DENDRAL, I emphasize the importance of a paradigm in which a part of knowledge is systematically embodied in a program and thereby "packaged" for use by a certain kind of client in a certain kind of situation. But there is another, diametrically opposed paradigm that is perhaps even more important, one where every effort is made to avoid packaging. The objective is to provide a vast array of informational parts out of which the user can construct, in a simple way, almost anything he likes. This has been called the Tinker-Toy or the Erector-Set approach. The main current embodiment of this idea with which I am familiar is the Consistent System of the Cambridge Project.

The last service or system that I want to mention in this necessarily incomplete list is the one that will turn the network into a medium for artistic expression. Most people view computers as machines that calculate or as systems that retrieve stored information. I think that it is essential to appreciate a third view, that computers and programming constitute a medium in which people can create things.

What makes programming fascinating, of course, is that the programmer finds himself dealing with a fabulous medium in which he can create castles in the air and have his boss call them programs and pay him for them. Anyone who has had the experience of building a computer program model has some feeling for the importance of synthesis in bringing the computer into the scientific process. I think that, within a few decades at any rate, it will go well beyond programming and modeling. The computer network will turn out to be the perfect plastic medium, ideal for creating objects of art of some kinds with which we are now familiar and of kinds of which we have not yet conceived. Note that the facilities for modeling are in their infancy and that the facilities for other forms of art essentially do not yet exist.

But what about the users? One naturally projects into the future when he thinks about computer-communication networks. Yet the real world is full of real people who want to see a real network being used for real purposes. On the other hand, it is obvious that if one picked a real man off a real street and sat him down at a computer terminal, a terminal connected to any network in existence, no use would result. That, in a nutshell, is the problem. I have a few things to say about it, but I do not claim to have a solution that will please the impatient.

The network I know best, the ARPANET, is a joy to programmers and researchers who know enough about programming to find their way around and to master procedures that have not yet been fully mastered for them. It attracts creative programmers, and creative programmers create the kinds of systems and services that they like and need—systems and services that have much in common with those that other creative people, not programmers, like and need. My first suggestion, therefore, is that we be careful, in our effort to develop usership, not to kill the goose that lays the golden egg. The need for creative programmers is nowhere near its end. As I have tried to suggest by envisioning future network uses, it is actually only beginning.

Systems and services developed by creative programmers usually have to go through a process, not yet fully understood, of clarification, documentation, impedance matching, and perhaps repackaging. This process requires that there be a kind of marketplace in which nonprogrammer users can sample offerings, try to use them, complain, and hopefully be heard. I doubt, however, that this marketplace will serve its function if it is viewed primarily as a marketplace. It should be viewed primarily as a part of the process of developing systems and services and not as a part of marketing, sales, or profit making.

As I tried to indicate in my projections, most of which were over a decade or two, most of the services and systems that will make networks vital are in their infancy or have not yet been conceived. Therefore, the best strategy for any movement to attract users to networks seems to be to select users who will be creative and productive and will bring into being the services and systems that later users will require. This concept of a user as a developer of services or systems may not be comfortable, but most of us have seen how important it is to have their point of view play a role in development as well as that of the system designer and programmer. I propose not merely to bring the creative user into the picture, however, but to concentrate on him and to postpone, perhaps for a rather long time, the user who will not create through using. Most of the areas in which networks will eventually be vital do not provide a good enough array of services and systems to make the demanding user happy. It would be a mistake to bring many of them into the picture before it is ready for them.

I am not saying, however, that computer-communication networks as they stand are useless. Some uses are already sufficiently developed to attract customers and serve them effectively. The main such use, by and large, is access to computer resources that are not available locally. People are finding it increasingly advantageous to go through a network for routine computing. Moreover, on the immediate horizons are specialized computer facilities (for example, ILLIAC and the DATA-COMPUTER) that will open new dimensions of capability that can be reached through networks. Another use that is ready and important is quick mail. The way quick mail is catching on in the ARPA network suggests that it will be one of the greatest near-term attractions.

But networks as they stand will not support genuine use. The role of the universities in the development of the network concept is, in my judgment, to create new services and systems and dimensions of networking. There is a great deal to be done. What we must concentrate on now is the role of the user in the realization of potential.

Reference

1. Lawrence Halprin, "Humanizing the City Environment," *The American Way*, in-flight magazine of American Airlines, New York, N.Y. November issue, 1972.

6

ROBERT E. KAHN*
Department of Defense
Advanced Research Projects Agency

Status and Plans
for the ARPANET

The ARPANET is a computer communication system whose creation
involved pioneering efforts in computer resource sharing and "packet
switching," the transmission and routing of digital data in groupings
called "packets." Through the development of an interface message
processor (IMP) and the use of wide-band leased lines, a system was
constructed that provides reliable and virtually error-free packet-switched
wideband communications and enables computers at major United
States research centers to be connected effectively and economically.
I should like to describe that development briefly and to indicate the
directions in which recent work is headed to extend the network concept.
   During the first year of its development the ARPANET grew from
an experimental four-node network connecting the University of
California campuses at Los Angeles and Santa Barbara, Stanford Research Institute, and the University of Utah to a network of fifteen sites
across the United States. These sites were primarily centers of computer
science research, supported by ARPA, each of which had one or more
large time-sharing systems as well as other computer resources such
as digital storage facilities and special-purpose hardware. Previously the
normal mode of operation for these centers was essentially "self-
contained"; that is, each site arranged to obtain, in-house, the resources
it needed to satisfy its computing requirements. This often led to the
costly duplication of resources at sites across the country. The network
greatly affected this situation by offering the promise of access to and
sharing of resources otherwise uneconomical for any one site—much
less all of them—to obtain. Certain of these resources could now be
procured to service the larger user community. Other specialized resources that already existed elsewhere soon came into demand as well.
   With the increase in the number of computer resources available in
the network and with protocol development efforts producing usable
network software for the computers, a small amount of useful computer-
to-computer experimentation began. But the amount of traffic at that
time was quite small for several reasons. In the first place, users knew

* The opinions expressed here are those of the author and do not necessarily represent
the position of the Advanced Research Projects Agency.

little about the systems other than their own. Documentation was often unavailable or, at best, unclear. But more important, the mechanisms needed to make the systems capable of convenient and reliable performance were not developed. So the usage grew slowly at first, generated primarily by systems programmers and network designers. By the summer of 1971, traffic was typically being measured in terms of tens of thousands of packets per day.

In August 1971, we began to connect to the network user sites that had no computing resources of their own and that expected to draw on the powerful resources already available. A modification of the IMP called a terminal interface message processor (or TIP for short) was developed to service them. This user community, fully dependent upon the completion of the communication software and documentation to begin using the network, provided a strong stimulus to organize the network's operation and resources. The traffic now began to increase noticeably. By spring 1972, the network had grown to twenty-four nodes, and the traffic flow was approximately 600,000 packets per day. As of December 1972, the network contained more than thirty-five nodes, including a satellite link to Hawaii, and the traffic had increased to 1.3 million packets a day. This is approximately 13 percent of the network's capacity. At this rate of growth the current capacity of the network will be reached by summer 1973.

If this rate of growth continues, we will shortly have to begin increasing the network's capacity by ordering more communications circuits or by increasing the capacity of existing circuits. The beauty of the ARPANET is that it is an evolving structure, both in capacity and connectivity as well as in capability. The network is now beginning to play the role of a marketplace in that the users can exercise discrimination in selecting better quality services, more economic machines, and the like. In a completely free market, each resource on the net could be priced in accordance with user loads so as to recover their necessary costs. Presumably a competitive situation would arise that exhibited key aspects of the free marketplace. But the current situation departs somewhat from the ideal model. The initial fifteen-node communications costs are not yet passed along to the users directly; rather, ARPA provides the service to its contractors and charges only the incremental share of its costs to other government agencies. Furthermore, certain resources are provided on government-furnished equipment, while others are privately owned. By carefully accounting for the value of all resources, even these minor differences will be resolved as long as the capacity of each network resource is consistent with the user demand for it.

The development of the network's technology is continuing on several fronts. First, with regard to the ARPANET, a modular high-speed IMP is being constructed by Bolt Beranek and Newman to increase nodal reliability so that the access to any resource through the communications subnet will continue to remain as great as or greater than that of the resource itself. The higher speed, of course, will allow circuit speeds of over a megabit a second to be connected and the ultimate network capacity to be substantially increased.

Several countries, including Canada, France, and Britain, have independently embarked on the design or implementation of packet-switched networks. It will soon become necessary to provide for the interconnection of these networks to allow long-haul data communications between terminals and resources separated by continents. Satellite communication provides an effective means for achieving worldwide coverage, and for this purpose we are proceeding to develop special IMPs that use wide-band satellite circuits in a novel and economic way. That is, the "satellite IMPs" will work in cooperation with one another so as to share a single satellite circuit among themselves. This sharing is achievable because it is possible for each satellite ground station to transmit a signal to the satellite on the same channel (at different nonoverlapping times) and for the satellite to broadcast this signal to all ground stations within its coverage. This approach is expected to be tested later in the year.

Another effort that we have begun is in the extension of packet switching to radio communication in order to handle mobile terminals. Small hand-held terminals such as were described by Roberts in the 1972 Spring Joint Computer Conference[1] will certainly be included, as well as large vehicle-based terminals, sensors, and portable instrumentation of many kinds. This type of network can offer effective error-free mobile communications in urban and rural environments and can provide an attractive potential alternative to dial-up facilities for certain types of local data distribution.

In the next few years we expect to see a number of firms obtain common carrier licenses to provide commercial packet-switched communications. Many of these companies will undoubtedly apply technologies similar, if not identical, to those produced under ARPA research and development programs. Thus we see the effect of our efforts not only being directly relevant to the country's defense needs but also having an important impact on the commercial sector as well, as the benefits of packet switching are more fully understood and applied.

Reference

1. Lawrence G. Roberts, "Extensions of Packet Communications Technology to a Hand-Held Personal Terminal," *Networks for Higher Education,* Proceedings of the EDUCOM Spring 1972 Conference, EDUCOM, Princeton, N.J., pp. 295–298.

7

MAX P. BEERÉ
Tymshare, Inc.

Commercial Data Networks
Using Available
Common Carrier Facilities

The problem of getting data from here to there as needed, when needed, and without error is not one to be taken lightly. The solution differs with the case in question. It is a problem that will not stand still long enough to be measured for a solution.

When we speak of networks, it is easy to confuse the job of the network. It is simply a device to convey man's intellect, while the computer is a device to expand or multiply man's intellect, and the terminal is a device that changes the information of man's intellect into a form that can be conveyed to another destination. This paper deals with the specialized network solution needed to serve computers and terminals.

Since a network is merely a matrix of "pipes" designed to carry data from its source to its sink, our problem would seem to be one of only having to ensure that the size of the pipe is consistent with the need. To do this sizing job, one must first know the parameters of the need: how much goes where, whence it will come, and in what volume. The challenge is overwhelming, but do it we must.

Once the problem has been defined or bounded, it must be disassembled into its components: computers, software, network, and terminals. Each component must be addressed as a related part of the problem to be solved. Once each individual issue has been dealt with, the components are reassembled into the working system and tested or debugged. The next step is to expand the system to meet the need.

In short, the effort should be to bound the problem, disassemble it into component parts, solve individual issues, reassemble and debug, and expand the new system to fit the need. The objective of associating the user of computer power and the computer to which he needs access should be kept in mind throughout this exercise. The important link is not computer-to-computer, but terminal-to-computer, which greatly simplifies the problem: increase efficiency and expediency of use. A computer-to-computer network necessitates the use of wideband facilities of 50 kilobits or greater, which usually cause the computer to perform inefficiently and expensively. The terminal-to-computer network, however, works very efficiently over voice-grade facilities that are relatively inexpensive and readily available. True, with large groups of users there must be a high-usage trunk that could and should use wideband frequencies. So we must have a balanced, well-designed network that can be expanded or contracted in the most economical way to fit changing needs.

Let us look at some figures (Table 7.1) that more or less describe the general problem of computer communications. These figures were taken from a Frost and Sullivan report made in 1971.[1]

Table 7.1. Use of Communication Terminals

	Year			
	1965	1970	1975	1980
Number of Computer Systems in Use	31,000	81,000	175,000	250,000
Percentage of Computers with Terminals	6%	28%	52%	74%
Number of Data Terminals	35,000	280,000	850,000	2,600,000

Frankly, I feel that the percentage figures are very low because they do not include the Touchtone telephone, which is a viable data terminal and is going to become more and more prominent, especially in the area of education. This phone is easily available to students and is very inexpensive. Note that the number of computer systems in use are expected to increase by a factor of 8 between 1965 and 1980, while those associated with terminals will increase by a factor of 99. The terminals through this same period will increase by a factor of 74. This is a strong indication of the ground swell need that is developing for networks to tie the terminal and the computer together.

Our immediate problem is to put people in touch with computers, especially with computers distant from the general using group. Since computers seem to be located primarily in areas of population density, it seems readily apparent that the only available means of serving these computers with networks is via existing common carrier facilities.

Of course, specialized data networks can be built using available transmission equipment, but only if the system is small and not widely dispersed; otherwise, large amounts of capital and lengthy time intervals are required. I prefer to regard this method of accomplishing the data network as an extension of data service rather than as the answer to today's dilemma. The need for computer network service is now; therefore, we must use what is available now. We cannot wait for the special data service networks that are in the offing because it will take too long for them to expand to include all the areas that must be considered here. And, as they are advertised, these new networks do not relate to the total problem. The advertised decrease in potential errors by an order of magnitude is good, but not good enough—several orders of magnitude are needed. Slightly lower costs at some speeds are good, but not good enough—lower costs are needed across the board. The lack of potential for alternate routing in event of line or tower failure is unsettling to contemplate, to say the least. No one has suggested that his special data network will offer speed or code conversion. Will the network be insensitive to the many kinds and types of terminals, or must some be catered to at the expense of others? These are a few thoughts one must consider in planning for his short-term needs. In the long term, five to ten years from now, the situation could change appreciably, and probably will. But the answer is to build for today's needs, from what is available today, using the entrepreneural guts, diligence, and expertise to fashion this material into products needed to meet tomorrow's challenge. The United States has a long history of doing this. Let us just do what comes naturally.

Let us look at a progression of possible network answers to our computer communication needs. The common carrier communication network available today to provide the paths for terminals or computers to talk to computers is the same network that people use to converse with people. It is undoubtedly the world's most advanced network—a people-buffered network. I suppose it is also the world's greatest general-purpose data network, too. But it is not good enough. It goes where we need to go, which is everywhere, but its data frailties have to be overcome. Data systems are too sensitive to line noise and other irregularities to be satisfied with the message network as it now exists.

Rather than explain in detail each of the many different types or varieties of data networks in use today, let us describe different types of computer-oriented networks in order of increasing potential importance.

- Networks that emanate from computers using the factory-designed front end equipment that is integral to the computer and that allows a certain mix of terminals to reach the computer via private or exchange lines.

- Networks that employ a multiplexing scheme. This configuration allows certain economies of communication costs and line efficiencies.

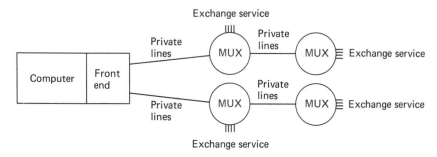

- Networks that use programmable computers as front end can connect many types of service to multiplexors, adding even greater flexibility and economy.

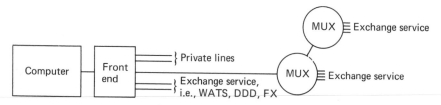

- Networks that use computer front ends and computer nodes in the net to permit error detection and correction, alternate routing, line load leveling, multiple terminal configuration, and so on. Again, added flexibility, economy, and efficiency result.

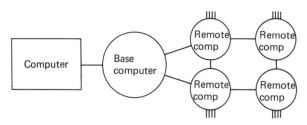

- Networks of computers that combine the potential of all the previous configurations, which allow the user to go wherever necessary to get the service required. This is an ultimate configuration that is perfectly consistent with today's state of the art.

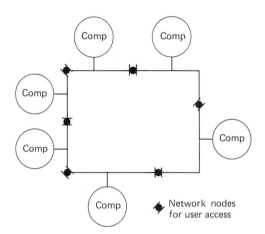

Network nodes
for user access

To accomplish progressive growth toward the last network configuration, we must overcome obstacles set up to protect and control communications. We must deal effectively with the regulatory agencies—OTP, FCC, PUC—and with the common carrier companies.

The addition of computers in the nodes of the network and as front ends to the host computer makes it possible to solve the computer communications riddle. The solution is, in fact, the evolution of a totally new service. This service has been given a name that both identifies and describes: Value Added Network Service, or VANS for short. It is upon this new service that we must depend for our near-term computer communications networks.

VANS, by definition, is a computer communications service based on common carrier facilities. This meets our previously mentioned requirement of using the available common carrier service to solve the data network problem. Basing a service on common carrier offerings alleviates a lot, but not all, of the potential regulatory problems. If you have an existing network for your own use, such as Tymshare does in TYMNET, it is permissible to allow joint use of the facility to other users who need service in the area your private lines may cover. The tariff is FCC 260, Section 2.1.1 (A) through (G) and Section 3.1.5. But allowing joint use is only part of the solution. The key to Value Added Network Service is, as the name implies, the *value added* to the garden-variety common carrier offerings. These values are the following.

Error Protection
The existing common carrier service has an error proneness of from

1×10^5 to 1×10^7, depending on the facilities, the time of day, and the bandwidth used (speed of data). To be of maximum value to the data transporter, protection from errors of several orders of magnitude must be available. This requirement calls for error protection features to be added to the common carrier service. There are two common varieties: forward acting, where the corrective information is carried with the data, and rearward acting, or correction by retransmission.

The forward-acting type is a highly complex form of correction and adds increasing redundancy as you increase the protection. Thus the line efficiency for information throughput goes down. Retransmission is by far the more generally used because of its relative simplicity. A checksum or parity check rides along with the block of data. If errors change the block, the change will be detected and the block will be re-transmitted automatically or on request of the receiving device. Using this type of system, it is possible to lower the error rate to a very accept-able 1×10^{12} to 1×10^{14}. TYMNET presently has an error prevention scheme that may allow one questionable bit in 4×10^{13} good bits.

Alternate Routing
Presently, the major carriers have the capability of alternate facilities in the event of catastrophic failure. The alternate routing can be arranged in as little time as three minutes or as long as three days. This is accept-able to voice traffic but not to the users of computer communications. Too much revenue is at stake, and too much high-cost equipment and technology are idled when the communications facilities are unre-sponsive. The need is for a computer communications network that em-bodies in its design automatic alternate routing in the event of facility or node failure. TYMNET has this capacity.

Line Load Leveling
This feature is related to alternate routing. If the primary route from a given node is approaching saturation, secondary routes must be avail-able and automatically useful. Maximum economy of facility use will result, with the big plus of maximum service.

Full Duplex Operation
To understand the virtues of full duplex operation, the user must use the service. Full duplex service allows the user to type ahead, that is, to continue typing into the computer while the computer is answering

the last query. It is possible to stop or modify a response from the computer at any time without waiting for the response to end. The computer-terminal discussion can go on in a parallel rather than in series. The common problem with this service is that the echo response from the computer is noticeable when the terminal is beyond a thousand miles from the computer. This is because of the time required for the line to transmit the character from keyboard to computer and back to the terminal to be printed. TYMSHARE has solved this problem by having the local node echo back the characters. In the event of simultaneous typing and receiving, a special protocol assures that all characters are fed back to the terminal in the proper sequence.

Speed and Code Conversion
The lack of terminal standardization, which allows for many different speeds and codes makes it mandatory that a general network scheme be devised that will be capable of handling any terminal with a minimum of setup functions. This job is not overly difficult if the nodes are software-oriented. Some of the more advanced hard-wired multiplexors now have this feature. In TYMNET, a terminal's code, speed, and carriage return characteristics are determined through the typing of one identifying character during the log-in sequence. The software needed to handle the particular terminal is called in to handle the interaction.

Multiplicity of Terminal Types
This feature is related to multiple code and speed in that the node must be able to accept calls from many kinds of terminals simultaneously.

Single Rotaries
To be most efficient on line usage, the lines associated with the individual nodes must be in single rotaries rather than very inefficient multiple rotaries, each catering to a particular type, speed, or code terminal. (Individual rotaries must, however, be assigned to different types of service, that is, WATS, FX, and so on.)
 Two networks in operation today are oriented around the foregoing considerations. One is ARPANET, a 50,000-BPS government research network, which is organized for the most part around a computer-to-computer technology but allows for terminal-to-computer activity. TYMNET (Figure 7.1) is organized around a terminal-to-computer protocol but has a limited computer-to-computer capability. All nodes in

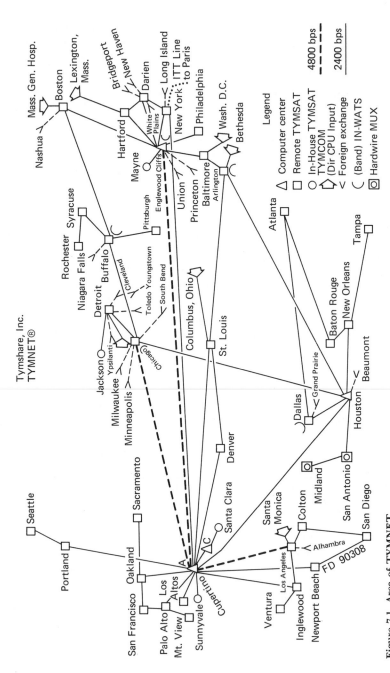

Figure 7.1. Area of TYMNET.

TYMNET are built around minicomputers. The network operates at a profit and fills the recognized needs of today. TYMNET will expand to meet new needs as they manifest themselves.

Several other commercial networks in operation today are beginning to evolve into Value Added Network Service. The United Computer Science Network, General Electric's time-sharing network, Control Data Corporation's Cybernet, and Computer Science Corporation's Infonet are the major ones. Since all these networks are based on common carrier service, it should be relatively easy to interconnect them when the time comes; but that is another story. TYMNET is used to disperse the computer power embodied in twenty-three XDS 940s and three DEC PDP 10 computers to our many customers here and in Europe. The network now has some 40,000 miles of private line facilities interconnecting some eighty-four nodes, called TYMSATs, covering the United States and extending into Europe.

A VAN service on TYMNET allows a customer to use the network to accomplish interaction with terminals that wish to gain access to their computer. They have the advantages of VAN service without the responsibility for establishing or maintaining the network. At present, these network customers have a variety of computers on TYMNET: IBM 370/155, PDP 9, IBM 360/55, Burroughs 6700, CDC 6600, and Burroughs 5500. The different kinds of computers are not able to talk to one another, but they have no reason to. All computers can be reached by their own families of authorized users at any time from anyplace TYMNET exists, which can be anyplace where common carrier facilities are available.

The medium is available to satisfy the needs of the widely dispersed user of nonlocal computer power. That medium is Value Added Network Service being offered by companies like Tymshare, Inc. The computer communications network solution is available. We must now expand that solution to fit the need—after we have bounded the need!

Reference

1. *Computer Data Terminals Market,* Frost and Sullivan, Inc., New York, N.Y., 1971, pp. 27–28.

8

NORMAN R. NIELSEN
Stanford University and
Wellsco Data Corporation

Network Computing

In considering the possibility of a new national network or in planning who should use networks for what purpose, it is appropriate to consider recent studies of network activities. Over the past four years the National Science Foundation has funded some twenty-five regional computing networks. A team consisting of Gerry Weeg and Jim Whiteley from the University of Iowa, Fred Weingarten from the Claremont Colleges, and the author of this paper spent more than a year looking into the operation of some of these networks.[1] This paper summarizes some of the problems and issues that were uncovered in the course of that study.

One should bear in mind that these regional computing networks were constructed primarily for the purpose of curriculum development or student education. Although some research use was made of these networks, it was small and was certainly not considered in the design. Further, the remote nodes on these networks for the most part did not have extensive computing capabilities or staff expertise. Hence, not all the findings of the study will apply directly to discussions of a national network. However, many of the issues that surfaced will be relevant.

Motivations

A general observation is that to be successful a network must benefit both the provider and the recipient of the service.

If there is not something in it for everyone, the network is likely to fail. Consider first some of the benefits that might be offered. From the standpoint of the central facility, possible motivations for network membership would include:

- Additional revenue. At the time that the NSF regional networks were started, many central facilities had deficits and excess capacity and hence, given the low marginal costs of incremental computing service, were looking at serving a regional computing network as a means of increasing marginal revenue.
- Facility expansion. The higher level of demand might provide the support base for new, larger, or in some sense better or more exciting equipment.
- Prestige. Involvement in this new activity might place the facility in the forefront in the eyes of others around the country.

Possible motivations for remote institutions to join a network would include:
- Service variety. A network can make available numerous languages, systems, packages, data bases, and so forth.
- Cost reductions. A central facility offers economies of scale in both hardware and system support.
- Communication. The network's existence provides an excuse for individuals at the using school to associate with those at the central facility, leading to the opening of communication on a number of curricular and technical subjects.
- Assistance or technical support. The reservoir of talented personnel at the central facility can be tapped as part of the regular network activity.

It is important to note that the preceding items are possible benefits that might result from an institution's seeking network membership. The word possible deserves emphasis since the benefits may or may not be (and oftentimes are not) realized in practice.

Central Facility Issues
The following are some of the issues facing management of a central facility.

1. Although network service can indeed provide a facility with incremental revenue, it seldom provides an incremental profit. Certainly in the short run a facility can sell raw computing capability at in incremental profit, since the marginal costs are so low. In the longer run, however, if the network is to be a success, the central facility must generally increase its supporting activities. This is manifested in software development (for special systems or packages needed by the using institutions), additional or revised documentation, consulting help, circuit riding, and so on. When all these additional costs are added into the calculations, the incremental profits tend to disappear rather quickly.

2. Network service can easily lead to the preoccupation of key staff members with network-related problems. Again, this is not something that need happen, but staff members in those networks which were successful tended to spend more time than budgeted on these problems. As a result there was not only an indirect subsidy of the network but also a reduction in staff attention to the problems of the host facility. Thus there is likely to be a real though unstated cost to the host institution.

3. Network service can result in a number of changes in the capabilities, service offerings, staffing levels, and even host institution service of the

central facility. As additional user-oriented individuals are brought in, the composition and focus of the central staff change. At times it may also be necessary to degrade service to the host institution in order to accommodate the needs of the remote user community.

4. A successful network operation may well lead to changes in facility ownership or control. The more successful networks tended to have some degree of remote institution control over the central computing facility.

Remote Institution Issues

A number of critical issues also faces the managements or administrations of the remote or member institutions in a regional computing network.

1. The overhead associated with the offering of many options and facilities may make the network an expensive means of serving small applications. A network facility, if it is truly general-purpose, has to serve a wide variety of needs. This leads to a number of items of overhead software (for example, command language, program supervisor, terminal interface, sophisticated data management system). Consequently, even a simple application may incur a fairly significant amount of overhead in getting to the right system, setting up the interface, and so on. As a consequence, processing that application on the network becomes rather expensive compared with other means of doing it.

2. The charging procedures of most networks result in a remote institution's accepting a liability for potentially very high charges. An institution with a small stand-alone computer generally pays a fixed monthly bill regardless of its usage. On the other hand, an institution using a network generally pays only for the services actually used. Potentially, however, a student or staff member can run up very large charges, since the network computer has (relatively speaking) an enormous capacity. Although there is technically an upper limit on what those charges could be, that limit is so high as to appear unlimited compared with the expected level. Hence, the institution must always be on guard against potential "bankruptcy" from unauthorized or excessive computer use.

3. The services offered by a network may not be appropriate for the intended user in terms of sophistication, complexity, ease of use, or mode. All too often a network is an extension of a central facility designed for the sophisticated user, making the use of these facilities difficult for the novice. Similarly there are user problems when a network

provides time-sharing service but not batch processing (and vice versa).

4. Difficulties may abound in the interface between the remote user and central facility personnel. Staff members at the central facility are generally used to dealing with a fairly sophisticated user population. Even though this population contains novices, they are surrounded by knowledgable people. This is seldom the case at the remote institutions. Thus there are often problems of materials and conversations being pitched at the wrong level, of central attitudes of superiority and inferiority, and the like.

5. Unreliable service may be a source of significant disruption and frustration to the using institution and its members. Poor computer service clearly affects any user, but the remote user, being more isolated and thus psychologically more dependent on the service, is likely to be affected more severely.

6. Reliance upon network service generally implies that an institution has given up some control of its computing capabilities to an outside body. But computing is becoming such an important part of an academic institution's operation that there is resistance to this loss of control and there is a strong pressure to regain it in some measure. This desire is reinforced when central facilities make unilateral decisions for the benefit of their institution that impair the ability of the remote institution to carry out its functions. Examples of such actions include changing service hours in the middle of an academic term (thereby eliminating service to an existing computer laboratory section), installing a new operating system during the school year (thereby making service a great deal more unreliable and severely affecting student project deadlines), and suspending remote service for a day (to serve an application of the central institution). If a network is to succeed, it must often accommodate this desire of the remote institution to have some control over the operation of the central facility. This control may take the form of membership on a user committee or board of advisers, or it may take the form of actual ownership of a share of the central facility.

Problems

The following review of operational problems by no means forms a complete catalog, but it does summarize and provide insight into some of the more important difficulties.

Network service is a very emotional subject. Psychological considerations are often of paramount importance. This is not to say that technical factors are unimportant; but at times they can certainly be of secondary importance. This is why it is possible for a network offering poor service to succeed while a network offering very good service fails. The following, then, are areas that must be addressed in the design or construction of a national network.

1. Capabilities
It goes without saying that a network should offer the types of services the user community believes desirable. This covers the range of batch services, data-base services, and time-sharing or conversational services. A prospective user will view a network that fails to offer a desired capability as seriously deficient or unsuitable regardless of its other features or facilities.

2. Key Personality
Network service often succeeds at an institution solely because someone believes that it should and will succeed. Conversely, network usage may well fail—in fact, is likely to fail—if an institution has no such person. The study team found that an influential person who is willing to push for the network can have a tremendous impact. With his presence the user community is likely to view the choice of network service as being right until proved wrong; otherwise the network choice is often viewed as wrong until proved right. Clearly, with all of the problems that arise in establishing a network and initiating service to an institution that joins it, things will not go smoothly in the beginning. The former attitude thus helps immeasurably in seeing the user community through the difficulties.

3. Self-fulfilling Prophecy
A network is often a self-fulfilling prophecy. That is, it is viable if that is the way the users see it and is not if they do not. If faculty members do not see the network as having a long-term future at their institution, they are very reluctant to invest their own time and energy into the development of programs and curricular materials to enable their students to take advantage of it. Without these materials usage does not develop, and without usage the network withers and dies. On the other hand, if the network is seen as viable, faculty members do invest time and effort. Materials are developed, and network usage also develops. This very activity causes the network to take root and succeed at that institution.

4. Aspiration Level

A very interesting observation of the study was that an institution's aspiration level often determines for what purpose it will use the network facilities. In other words, an institution will use a network as it thinks a network ought to be used. Thus, at one school only beginning programming exercises are run on the network because a belief runs through the staff and administration that network computing means the processing of small student programs in support of programming classes. At another institution on the same network the computing capabilities may be used in support of physical science laboratories, English literature classes, geography, and a variety of social science courses. Thus, although a network may be designed to fulfill a particular requirement or serve a particular category of needs, its actual application may turn out to be quite different if the using institutions do not internalize these objectives.

5. Perspective

A tremendous difference exists between the outlooks of the central facility and the remote institutions on any given issue. Thus, while the central facility may feel that a particular action is good and in the best interests of the remote users, those users will perceive that action as bad and not in their best interests. There is also a role perception problem in that each group has different images of the roles of the central facility and a remote node. Thus, there is a serious problem in determining just how a network should operate. It is very interesting to note, however, that such differences in perception are a function not of the institution but of its role on the network. When a remote institution drops out of a network, sets up its own facility, and attempts to serve other institutions as a central facility, it very rapidly adopts the "inappropriate central facility viewpoint" it had criticized only a short time before.

6. Motherhood, Apple Pie, and Networks

An opinion study carried out at the University of North Dakota and confirmed by the study team's investigations indicated that nearly everyone was in favor of networking. It is, in a sense, the in thing. Few persons, however, expected to be served by a network; nearly everyone assumed that his facility would be a supplier rather than a user of service. The prevalence of this attitude is indicative of some of the psychological barriers facing those seeking to initiate network operation.

7. Empire Building

Everyone has heard about the individual who is more concerned with building up his own empire than with anything else. Yet the pervasiveness of this type of attitude among the participants in the regional network program was quite surprising. In nearly every network studied, one or more schools dropped out and attempted to establish their own networks. Now it is certainly true that there are situations where it makes a great deal of sense for a school to provide its own computing facilities. But the schools dropping out were not generally in this category. Their motivation for starting another network focused on using it to support a larger and more powerful central facility for themselves.

8. Distorted Comparisons

It is often the case that a school will have some computational capabilities of its own as well as a connection with a network. In such instances the comparisons between the network service and the in-house service are likely to be quite distorted, with a bias in favor of the local facility. This is not to imply that all such comparisons are improper or that all of the errors to be described now will be committed. Nevertheless, the extent of the distortions was surprising, and the following gives an indication of the types of problems that a network faces when it competes with a local facility.

a. Use of inappropriate terminal equipment. A member school may select a teletype as the interface with the network and a card reader as the interface with its own facility. For an institution oriented toward batch processing, the network services will be perceived as poor, slow, inappropriate, a step backward. Thus a local administration can have a significant influence on the perceived value of a network through its choice of terminals.

b. Self-imposed limitations on network services. Very often the service considerations for the local facility and for the network are not the same. For example, the terminal room for the network may be closed every day at six P.M. while the local computer facility may be staffed until midnight.

c. Differential performance standards. Even if a network and a local facility had identical patterns of downtime each day, the local facility would be perceived as offering a superior service. That is to say, network failures are considered intolerable, whereas local failures can be tolerated. The attitude toward the in-house facility is definitely supportive. The user is more willing to forgive its faults. The result is that his impression of the local system's performance is more favorable. Thus, network service has an additional handicap to overcome.

d. Differential accounting procedures. The typical local facility is operated at an essentially fixed cost per month regardless of usage. Thus it is usually made available to the user community as a free good, with waiting lines, long turnaround, or other de facto procedures used to allocate the resource. Hence, the local computer is perceived as being cheap. On the other hand, recovery for use of a network is rarely on the basis of a fixed price. Usage is the common determinant. Any usage whatsoever thus incurs additional expenses, and specific budgetary control mechanisms must be established. Use of the network is consequently perceived as being expensive even though total network charges may be less than the cost of supporting a local facility to provide the same services.

e. Biased cost studies. The charges paid for network service are generally what might be termed fully costed. That is, they include not only the cost of the computer itself but also the costs of space, utilities, and staff. The expenses for running a local facility are generally not handled in this fashion. The director of the local facility often has another position in the organization and manages the computer facility on released time. His salary is unlikely to be charged to the computer facility, however, since it is fully covered by his regular position. Student assistants may be employed by a state-supported work-study program and their wages charged to budgets other than the computer facility's. Similarly, space and utilities may not be charged to the local operation. Thus, the cost pool seen for the local facility is usually much smaller than it would be for a fully costed operation. Hence, network service faces an additional handicap.

f. Biased usage studies. The local facility and the networks are often employed for entirely different purposes. The local facility, for example, may be used to support the small compiling and debugging runs for students in a beginning computing course while the network facility may be used to support term project work of advanced students. This may be a good distribution of the work load, but if the total costs are accumulated at the end of the semester and divided by the number of students to compare systems, the local computer might come out with a cost of something like $50 per student while the network's record might be $150 per student. These costs do not, however, support any implication that network computing is less efficient or three times as expensive as the local system.

9. Support of Central Services

Time after time the study team observed the importance of the central facility's support for the remote institutions. This support encompasses developing materials, testing materials, converting packages to run on the network, consulting, training, and what might be termed "evangelical" activities. The existence of such support often differentiated the networks that flourished from those that did not. Although users wanted more of these services (no matter what the support level, more was always desired), the institutions were generally reluctant to pay for additional services. This finding raises a long-run question about how this type of activity can be supported. (The North Carolina Educational Computing Services' surcharge on Triangle University Computing Center's base rates is a notable counterexample.) Clearly, additional support involves additional costs and requires additional funding. But if users are unwilling to support these services that appear necessary for the health of the network, the long-term viability of such networks is questionable.

10. Chicken-and-Egg Problems

Starting a "good" network operation is very much a chicken-and-egg situation. Remote or member institutions do not wish to commit to a level of usage until they have seen how it will work in practice. By the same token the central facility does not want to invest in the developments needed for a "good" network (staff buildup, documentation, new systems, and so on) unless there will be a sufficient usage base from which to recover the costs. Yet without that investment the user institutions are unlikely to remain members or to make significant use of the network. Thus, each party may be viewed as standing around waiting for the other to make the first move.

11. Self-help

The existence of a network tends to make users much less self-reliant and more dependent upon others. This was a surprising and disturbing finding. A local facility is frequently run on a shoestring. It is up to the users to do things for themselves. Clearly some will not do this and will not become computer users, but many do. They seek out applications, they get things running, they "fix up" the documentation, and the like. On the other hand, a network generally implies a certain level of overhead organization—a network coordinator, a network interface at the central facility, a circuit-riding consultant. With all this aid, a surprising number of faculty members sit back and wait for things to be done for them. Nor is it just a matter of waiting to be told about the existence of a certain package, the attitude extends to mounting the package on the system, debugging it, tying it into the texts used in the classroom, and so forth. Clearly, this development of complete curriculum packages is quite expensive, and a network is unlikely to provide it economically.

This discussion is an attempt to illustrate the problems that should be addressed in planning for a national computing network. Of necessity only a summary of some of the findings of the study is presented. Anyone becoming seriously involved in networking is urged to read the study team's full report, "A Study of Regional Computer Networks," University of Iowa, 1972.[1] This report not only elaborates on the items discussed earlier but also goes into other areas of network operation. In particular, fifty-five different factors are discussed that are related to the provision and use of network services. Although no single factor is a necessity for a successful network, they are important in toto. Thus, anyone considering a network should find it helpful to evaluate his plans against these factors.

Reference

1. Fred W. Weingarten, Norman R. Nielsen, James R. Whiteley, and Gerard P. Weeg, "A Study of Regional Computer Networks," The University of Iowa, Iowa City, September 1972.

9

ROBERT L. ASHENHURST
Institute for Computer Research
The University of Chicago

Hierarchical Computing

The literature on hierarchical computing consists mainly of descriptions of specific systems, which are asserted to be (and presumably seem to be) hierarchically organized, and a few general discussions of some of the potentialities of hierarchical organization. A good example of the former type is a paper by Peter Lykos[1] and of the latter is a summary of a panel discussion led by William Bossert.[2] Some of the financial and management issues are discussed in the context of computer networks in a paper by Ruth Davis.[3] That all the three papers cited appear in EDUCOM conference proceedings is not to say that no papers have appeared elsewhere but rather that hierarchical computing, along with networking, has been a particular focus in the recent considerations of how computing may be made more cost-effective for educational and research organizations.

In this paper an attempt is made to delve a little more deeply into the attributes of computer and information systems that make hierarchical computing attractive, particularly in the context of computer networks. In the first part of the paper a general context for hierarchical analysis is outlined as a background for discussion in the second part of a specific example, the Minicomputer Interfacing Support System (MISS) project at the University of Chicago.

Concepts

The word hierarchy is used in general in various distinguishable, albeit often interrelated, senses. To impart meaning to the term "hierarchical computing" one must observe certain distinctions and likewise pay some attention to the various uses of the word system in the computing context.

Hierarchy and System

Herbert Simon in *The Sciences of the Artificial*[4] (Chapter 4) defines a hierarchic system as "a system that is composed of interrelated sub-systems, each of the latter being, in turn, hierarchic in structure until we reach some lowest level of elementary subsystem." He subsequently notes that the word hierarchy has a narrower meaning in its etymological derivation, "to refer to a complex system in which each of the subsystems is subordinated by an authority relation to the system it belongs to."

He proposes to use formal hierarchy for this more specialized concept.

The concepts of system and subsystem themselves, of course, require considerable elaboration. Suffice it to say here that the system-subsystem hierarchy can be defined in a variety of ways for entities found in the real world, and the way in which a system is "composed of" its subsystems varies in the different definitions. To see the need for admitting a variety of system decompositions for a given entity, one has only to think first of a human being as a system composed of skeletal, muscular, and glandular subsystems and then of the head, trunk, arms, legs, and so on as subsystems. A useful notion in almost all system situations in Simon's concept of a system partitioning the world into an inner environment and an outer environment[4] (Chapter 1) and providing an interface between them. A hierarchy of systems and subsystems, then, also provides a hierarchy of nested inner environments, and often the outer environment for a given subsystem need be viewed as extending only as far as the boundary of the next higher subsystem.

Computer Systems

A fairly well delineated example of the system concept is that of the computer system, viewed essentially as a complex of hardware and operating system software. Hardware modules are classifiable by function as processor, memory, and peripheral, the main types of the last being mass storage, input-output, and remote transmission units. Various auxiliary types of modules having a connection or direct control function must be identified in a detailed consideration. Operating system software is also generally organized modularly, and although the exact position of the system boundary may be more open to question here, the criterion that the operating system consists of those programs and data sets whose "subject matter" and "content" concern the computer system itself and not the world around it provides a useful line of demarcation. The computer system as a whole, then, is an entity that can efficiently perform tasks, defined by programs and data sets, whose interpretation is arbitrary. The interface between the inner and outer environment for a computer system is defined by its input-output units, its mass storage units as they permit mounting and demounting of volumes of data (that is, tape reels or disc packs), and its remote transmission units insofar as they are regarded as communicating across the system boundary.

Given this definition, suitably refined, a variety of hierarchies can be discerned in the organization of a computer system. A complex computer system can be partitioned into subsystems that are themselves computer systems, and these may be interconnected to form a hierarchical array, a hardware configuration hierarchy. Note that this is a derived structure that may or may not be treelike and is distinct from the basic system-subsystem hierarchy, which simply aggregates hardware modules into larger and larger groups. To see the distinction, consider the basic example of a large central processing system fed by several smaller systems functioning as remote job-entry units. The hardware configuration hierarchy is then the treelike structure obtained by imagining the central processing system at the top of a diagram and the remote systems subordinate to it. It is a formal hierarchy in Simon's sense if the term authority relation is loosely defined. The system-subsystem hierarchy in this case, however, consists of the system as a whole at the top level and its various subsystems, one of which is the central processing system, at the next. The subsystems, particularly the central sytsem, can be further broken down into constituent subsystems to give more levels to the hierarchy.

This notion of a hardware configuration hierarchy is important because it is probably the interpretation most would apply to the term "hierarchical computing." Indeed, distributed "star" or "star-cascade" configurations are typical examples of computer networks exhibiting a hierarchical configuration. In the general network context the organization need not be hierarchical in the configuration sense, and yet the notion of hierarchical computing levels may still apply.

Analogous to a hardware configuration hierarchy is a software configuration hierarchy, defined by the relation of intermodule calls over a set of software modules. Again, a given set of such modules can be but need not be hierarchically configured because of exotic mechanisms such as recursive subroutines, coroutines, and the like. But in the most straightforward case—a software system organized as a main program and a set of subroutines, with each subroutine invoked by a call from a higher-level routine to which it eventually returns control—the software configuration is hierarchical.

Other hierarchies often arise in association with computer systems. A hierarchical file organization imposed on a data set permits a straightforward scheme of specifying and gaining access to individual data items. Operating systems are increasingly being organized so that they run as a set of processes—a process being defined roughly as a program-in-execution—with a hierarchical form of process control (each process has been invoked by some unique process, its "immediate ancestor") serving to maintain integrity of operation (that is, to avoid or escape from inter-process deadlocks and the like).

The sense of hierarchical computing that seems most fruitful for studying applications of computer networks is somewhat broader than those discussed so far. It is most conveniently set forth in the context of information systems.

Information Systems

A precise definition of an information system is not easy to come by, but nevertheless the concept is becoming increasingly well delineated. An information system may be thought of as a set of procedures carried out by people to meet some systematic need, generally associated with an organizational effort. Although the notion of computer systems is not inherent in this characterization, in practice they are a necessary adjunct to large and complex information systems. This leads to the concept of an information-processing system, a collection of programs and data sets that run on a computer system in support of and as part of the procedures of an information system. These programs and data sets are distinguished from those of the computer operating system by the direction of their "subject matter" and "content." In fact, in these terms the computer system is simply a specialized information system whose purpose is to implement a more general information-processing system within an information system.

A given computer system, however, may be assigned to serve in this capacity for several essentially independent information systems. It may also be called on to provide general computational service. This diversity of function can lead to problems in the operation of an information-processing system.

The distinction between an information-processing system and an arbitrary collection of vaguely related programs and data sets arises from a requirement that the information system of which it is a part be persistent and pervasive. Persistence implies that the information system will function over a substantial time span, perhaps indefinitely, continuing to satisfy the general needs for which it was developed. This requires the functioning of the information system to be routine and predictable but flexible enough to meet changing patterns of needs. Pervasiveness implies that the performance of the information system will be consistent with the human abilities and limitations of a large number of people interacting with it or affected by it. In particular, it should be assumed by the developers of the information system that this performance will not be dependent in any way on their continued personal presence.

The requirements of persistence and pervasiveness impose constraints on the development of information systems that make it a substantially more exacting activity than mere "applications programming." The ACM Curriculum Committee on Computer Education for Management has recently published recommendation for graduate professional programs in information systems development.[5] In this report general information system requirements are characterized in terms of six "abilities": capability (to fulfill the intended purpose), stability, modifiability, usability, operability, and maintainability[5] (Sec. 2). These attributes can be related to the roles of several identifiable groups of people connected with the information system's operation. First there are three groups identifiable at the "functional" level: benefiters, for whose activities the information system provides support; overseers, who have the responsibility for the satisfactory performance of the information system in some aspect; and modifiers, who are charged with specifying changes in the information system as requested by the benefiters or overseers. Besides these are three groups at the "operational" level: users, who interact directly with the system as agents for the benefiters; operators, who keep the system functioning in routine operation; and maintainers who incorporate the changes in the information system specified by the modifiers.

The modification function involves the tools of information analysis and system design that are applied at the specification stage of new system development, and the modifiers may be part of a systems development group. The maintenance function has more of a programming nature, and the maintainers may be those programmers involved in the implementation stage of new system development. The operators are generally those whose basic task is just to operate the computer system, and the maintainers may also perform a similar function ("operating system maintenance"). The overseers who are administrators of the information system may have special functions (such as "data base administration"), and collectively their function is different from that of the administration of the computer system ("computing center administration"). Finally, it is important to recognize that there may be people interacting with the information system who do not belong to the organization or department that runs the information system, and their role must be categorized. A case in point is the general public in the case of an information system built into some sort of a consumer service such as billing, reservations, or subscriptions. It is appropriate to ask whether the consumer is a beneficiary or a coerced user of the information system and presumably to try to eliminate the invidious implications of the latter role.

A functioning information system consists of a user interacting with one or more processes, defined in the technical sense of a program in execution, with associated data sets. The data sets are in the nature of "internal files" for the process, and they may or may not represent data bases, defined as a continuing representation of some actual collection, which needs to be updated as the real-world situation changes. The user furnishes input to the process, which may consist of both commands and input values. The user also expects output from the process in the form of text, tabular values, or display. In order for this interaction to work effectively, the system must provide the user a means for activating selected processes, and this requires that copies of the appropriate programs and data sets be available. There must also be provision for the user to introduce input and receive output. The basic interaction described, however, permits various patterns of control between user and process. The resources provided by the system to the user are also subject to various patterns of sharing and distribution among other users and among operators and maintainers, who play a similar interactive role when performing their functions.

Information systems described in this way furnish the context for the analysis of hierarchical computing possibilities. Many of the advantages of hierarchical system organization are mainly apparent when considering the specialized needs of information systems from the point of view of all who interact with them.

Determinants of Hierarchical Organization

The basic activity in an information system, then, involves the interaction of a user, one or more programs, one or more data sets, and some computer system elements such as an input-output station and a processor-memory combination. What is meant by program and data set in this context are actual copies on some storage medium of the symbolic entities involved. The most natural place for such copies to exist is on mass storage media in such a form that they can be automatically invoked in process activation. The input provided by the user, and the output acquired, are both mediated by the input-output station.

All the components, both hardware and software, must have physical locations, and those locations must be contiguous or connected by remote transmission equipment. In addition, the sites where the possible users may exist imposes some spatial constraints on the location of components, at least of input-output stations. The specifications for an information system generally include various response times, which impose temporal constraints on the interaction of user and process. But within these constraints there is generally considerable latitude for system organization, including various patterns of control and physical ownership of resources. A good solution requires consideration of aggregates of programs and data sets, aggregates of users (including in this category operators and maintainers), and aggregates of computer system components. A hierarchical system organization is generally attractive when many users are widely dispersed geographically and the user sites have fairly severe response time requirements. But it is also attractive with an identifiable set of processes that can be shared in common under less severe spatial and temporal constraints on the user-process interaction.

For a more detailed analysis, considerations such as whether two programs "do the same thing" and whether two users may be expected to "cooperate" or not become important. Without trying to put too fine a point on it, let it be said of two programs, data sets, or computer system components that they are identical, similar, or different. Identical programs and data sets are symbolically the same; similar programs and data sets "do the same thing" or "say the same thing" but in different programming languages or data structures. Identical computer system components are physical duplicates of each other; similar ones merely have general functional similarities. Similarly, we can say of two users that they are coordinated, comotivated, or independent. Coordinated users carry on similar activities or perform subtasks of a particular task; comotivated users merely belong to a group that is cohesive enough so that its members can be expected to have the same pattern of use. Obviously, the distinctions are subject to shades of meaning, but the object here is to attributes of general force even if not precisely definable.

The original idea of a "general-purpose" computing facility was to serve a variety of independent users, each responsible for a personal collection of different programs and data sets. Coordinated groups of users could work out their own arrangements for sharing programs and data sets of mutual interest. As the sophistication of operating systems developed classes of comotivated users were identified, and system operation was "tuned" so that the needs of the various classes could be better served. When job processing was the main mode of operation, turnaround time was the prime temporal constraint. All components were centralized, and the user had to carry his programs and data sets to the central site to run jobs. Further system development led to the possibility of off-site input-output stations communicating with the central site, permitting remote job entry. The time-sharing mode, permitting interactive access from remote terminals, made possible a fundamentally different type of operation. The introduction of multiprogramming and multiprocessing techniques on various levels meanwhile led to the more efficient deployment of system components at the central site.

Although these advances imply that in principle one giant centralized facility with dispersed input-output stations could serve all information-processing needs, this solution is unsatisfactory in practice. Aggregates of users desiring an elementary level of interactive service can apparently be served more satisfactorily by a computer system devoted to time sharing alone. The advent of the minicomputer has made the concept of small dispersed computer systems attractive for some applications, even those that may as a whole require several minicomputers' worth of computing capacity.

To understand this fragmentation and to see how hierarchical organization may offer even more satisfactory solutions to system problems, it is necessary to look more deeply at the needs of coordinated and comotivated users and the implications of these needs for the question of identity versus similarity of programs, data sets, and system components. Questions of control patterns and custodial arrangements covering computer subsystems also enter. Fundamental to the consideration are means for facilitating the sharing of programs, data sets, and system components when desired while preventing unwanted sharing. The most natural way of sharing a given resource is by multiple utilization. But replication is also possible, and the same effect can be achieved by imitation or "emulation" of one resource by a similar one. The other side of the coin is dedication of some resource, and this is enhanced by its isolation or insulation within the system of which it is a part.

All these aspects are found in the contemporary version of the general-purpose computing facility. Multiple utilization is enhanced by multiprogrammed operation, so that users' jobs can be interleaved on a single processor. Although input-output processing could be handled on the same processor, a separate subsystem, a "front end" processor physically isolated from the main system, is often dedicated to this purpose. The operating system in the main processor is insulated from the users' processes, as they are from each other, to prevent unwanted interference. Replication of processing components can also be desirable to decrease multiprogramming overhead and to permit greater flexibility in response to user demands. This brings about various multiprocessing configurations. Imitation capabilities make it possible to handle users' programs and data sets in a more versatile way, for example, by defining "virtual processors" with various capabilities.

The opportunities for applying these concepts become sharper for computer systems implementing more specialized information systems for coordinated or comotivated users. The second part of this paper illustrates this point in terms of a particular example, a hierarchical system being developed to serve the needs of scientific experimenters.

Hierarchical Organization for Laboratory Automation

Computers are increasingly becoming an integral part of laboratory instrumentation, and various systems approaches have been taken to making their use in this application more routine. The earliest form of computer support was to run many experiments off a single time-shared control system. The availability of minicomputers makes it possible in many cases to devote a minicomputer to each experiment. The difficulty of auxiliary tasks such as program development and debugging when performed on the minimal configuration of laboratory minicomputers leads to consideration of various ways to supply them with additional support. Many such systems have been developed, and several projects are working to develop laboratory support systems along more general lines. One of these projects is being carried out under the direction of the author and is used here as a specific example of hierarchical organization in application.

The MISS Facility

This project to develop a Minicomputer Interfacing Support System (MISS) is currently under way in the Institute for Computer Research at the University of Chicago.[6] It is one of several projects on hierarchical computing as applied to laboratory automation funded by the National Science Foundation. Since the facility is a good example of an application of the concepts of hierarchical organization of computer systems and information systems set forth earlier, it warrants a fairly full description here.

The MISS facility is structured as a trilevel support hierarchy supplying a unified set of capabilities to the scientific investigator but with a particular pattern of separation and sharing of support functions on the various levels. Level 0 is occupied by minicomputers, each dedicated to a particular experimental task (or coordinated set of tasks) and under control of its owner (or owners). Most of the programming is done by the owners and their agents (the scientific investigators and their assistants), and the owners are free to detach and reattach their minicomputers at will. Provision is also made for remote peripherals serving a group of investigators with several minicomputer sites in the same vicinity. These would be under the operational control of the users at the site and hence also form a part of level 0.

Level 1 is a system called the Minicomputer Operation Monitor (MOM). It provides on demand an array of standardized support services for level 0. These services include on-line peripheral handlers, a file system of limited capacity, some components of a "virtual operating system" for the minicomputers, and access to higher-level systems. All minicomputer users share the use of MOM, and it is important that it provide stable and reliable operation over extended periods. It is also desired that its running characteristics be predictable, permitting the operators to control the parameters affecting its performance. For these reasons level 1 is implemented on a dedicated computer system that is not available to run user programs.

Level 2 is a system providing higher-level supporting services to MISS users. It will be implemented as a computer network embracing a variety of host systems. Its services are considered in two groups: those provided by a system called the Data and Algorithm Distributor (DAD), and other supporting services. The DAD services form a package that is defined and documented by the MISS development group. They are available in a prescribed way through the connection to MOM, and users need not know what kind of a facility is used to provide them. Other supporting services can be developed by users, perhaps with assistance from the MISS group, for their own specialized purposes. Although the connection to this higher level is also through MOM, users wishing to exploit these other supporting services do need to be aware of the nature of the higher-level facility.

The DAD services in particular, and perhaps the others, can be distinguished as those used to aid in program preparation and other preliminaries and those used in connection with experimental runs "in production." Prominent in the former group are compiling and assembly services, which can be supplied in a fast turn-around batch mode. The latter group requires a greater variety of routines, some with response times that make batch service infeasible.

This way of organizing the system makes it natural that the "overall control" of the multiuser operation reside in MOM. The minicomputer sites are too dispersed to make it desirable for control to reside there, and in any case a prime purpose of the system is to relieve the scientific investigator and assistants of the need for knowledge and competence regarding the general system operation. The higher level, on the other hand, is inappropriate for the control site because it is a general-purpose computing facility shared both by MISS users and a larger user community whose interests are outside the MISS project.

The primary consideration in specifying level 2 as a network is the need to supply a versatile and flexible set of higher-level support services. A secondary consideration is the possibility this arrangement affords for replicating MISS at other network sites. Since DAD and the other support services are available symmetrically at any network node, effective replication at any node is achieved by duplicating MOM (hardware and software) there. A separate connection of some higher-level system to MOM, bypassing the network interface, would make the access of MOM to DAD "special" to the prototype installation, and the replication feature would be lost or at least rendered substantially more expensive.

A third consideration concerns the system developers, the MISS group. Level 1, the MOM system, is necessarily their access to the whole system. During the development phase it must be flexibly expandable and must incorporate new components easily. This flexibility is made possible partly by the nature of the MOM operating system, but it also depends on having a higher level of versatile and comprehensive system support functions. The same group must develop the DAD services in the context of a running MOM system. It is hoped that existing software can be the source of many of the programs above the basic operating system level for both MOM and DAD. A network making available a variety of processors and processing modes provides an ideal environment for this kind of development activity. Finally, the provision of other support services will necessarily be dependent on already existing programs. If these are usable in their original form on appropriate host computers, the work of the MISS group in making them available can be much facilitated. Developing the competence to make network resources readily available to MISS users is one of the important responsibilities of the MISS development group.

Higher-Level Support Analysis

The needs of research users for direct support of experimentation may be classified into five categories: (1) data collection, (2) automatic control, (3) human-mediated control, (4) auxiliary analysis, and (5) follow-up activity. Of these, 1, 2, and possibly 3 are essentially taken care of by the inclusion of a minicomputer in the instrumentation. Note, however, that data collection may have to be augmented by a bulk transfer function for large aggregates of data collected over time, and this is one of the purposes for which a higher-level system is introduced. The needs for 4 and 5 usually exceed, perhaps by a great deal, the capacity of the minicomputer configuration needed merely for data collection and automatic control. Thus, further processing power must be available. The five support categories are only those directly concerned with the production run and do not include program preparation and system debugging, whose demands for higher-level facilities have already been discussed.

"Auxiliary analysis" means the evaluation of models related to the experimental phenomenon while the experiment is actually in progress to aid in adjusting the apparatus ("human-mediated control") and in deciding whether the experiment is progressing satisfactorily and will yield valid and useful data. Such analysis may take the form of an abbreviated version of the postprocessing routines (for example, evaluation on the basis of a few points), a comparison based on access to large tables, or other specialized forms. The use of graphics for displaying the results of on-line analysis is becoming increasingly important. To support graphics effectively requires a mixture of lower-level and higher-level programs, and it seems that a hierarchical configuration is particularly suitable for this application.

Not all experiments can make use of a facility for auxiliary analysis on line, but many can, including several in the areas of current interest at laboratories where computer support systems are under development. Certain types of experiments using spectrographic and x-ray diffraction techniques, nuclear magnetic resonance, and the like are in this category. The on-line analysis required may involve fast Fourier transform evaluation, inversion of quantum mechanical matrices, curve fitting, and other standard types of model calculations that exceed the capacity of the minicomputer controlling the experiment.

The last term, follow-up activity, means the data reduction, model evaluation, and the like that normally follow an experiment. It is clearly desirable to have a convenient way of transferring the data collected in the experiment to whatever larger facility performs this postprocessing. This function can be merged with the above-mentioned one of bulk transfer of aggregate data. If the higher-level system is a network, data transfer and follow-up activity can be combined, making use of whatever remote host or hosts have the appropriate programs for postprocessing.

Both the on-line analysis and the follow-up activity require convenient and direct access to larger facilities. The effectiveness is crucially determined not merely by such access but by the likelihood that existing programs available on the facilities can be readily adapted to the particular application, and "general-purpose" analysis tools such as versatile language packages. In addition, for on-line analysis, there is a need for interactive and time-sharing running of moderate-sized jobs to give adequate response within the time constraints imposed by the experimental situation.

Hierarchical Analysis of MISS

Many of the design decisions for MISS are based on aspects of hierarchical organization discussed earlier. Here a brief outline is sketched, relating the levels 0, 1, and 2 characterized in the preceding section to aspects of users (including the MISS group in its role as users for development) of computer systems, and of programs and data sets. The analysis assumes that, in addition to the multiplicity of minicomputer systems served by one MOM system, there may be a multiplicity of MOM's served by a single DAD network system, even though the MISS project in fact only provides for one MOM on the University of Chicago campus.

A. Users:

level 0—The scientific investigators and assistants who are the primary benefiters in MISS. They are assumed to be prepared to develop their own programs for their applications that tend to change on a scale of weeks or months. But they are unable to afford extensive dedicated configurations in their experimental setup, and in fact they often have multiple uses for their minicomputers on their own or in collaboration with other investigators.

level 1—The MISS group, developers and modifiers of MISS, and programmers, maintainers, and operators as well, utilizing the MOM equipment and operating systems for development work on the remainder of MOM, and also for development of DAD network services, possibly in collaboration with groups associated with other MOM's.

level 2—No users at level 2 (not mediated through level 1) identified at present, although there could be overseers for the whole group of MOM's making use of the network for collecting MISS statistics and the like.

B. Computer Systems:
level 0—Minimally configured minicomputers, not necessarily all of one kind (similar, not identical), running a standard interface program to communicate with MOM, otherwise under control of an investigator for use in experiments.

level 1—Medium-scale hardware-software systems (MOM), dedicated to application (isolated from rest of facility), replicated at each site where MISS exists, providing minicomputer users with services of four kinds: (1) remote operating system for minicomputers, (2) limited-capacity file system for data needed on-line, (3) peripheral control system, allowing access to a variety of input-output devices, some of which can be (permanently or temporarily) at minicomputer sites, and (4) access to level 2.

level 2—A variety of host systems on an extensive network, providing fast turnaround batch and interactive capabilities of moderate response characteristics, including the DAD software, as well as other programs and data sets useful in the laboratory support context, all available symmetrically to any MOM system.

C. Programs:
level 0—Minicomputer programs developed for each application, stored and activated for use through MOM, and a minimal resident program, of standard specifications, for communication with MOM.

level 1—A standard set of programs, changing only because of overall system improvements, engineered for maximum efficiency in supplying MOM services, replicated in each MOM.

level 2—A standard set of programs (DAD), each needing to be resident at only one host site (but possibly replicated for backup), and some additional programs of special interest to users, all available symmetrically to all MOM's.

D. Data Sets:
level 0—Data sets (and on-line file space) directly associated with minicomputers only in special cases where rapid access during experimental run is needed, since otherwise access through MOM is sufficient.

level 1—A limited volume of data sets resident only in MOM, since MOM acts mainly as a buffer between the minicomputer and data sets stored on the network.

level 2—An extensive volume of data sets resident on the network, available through MOM via a host-to-host protocol, and permitting sharing by all MISS users equally without need for duplicate copies.

A primary consideration in this specification is the existence of the MISS group which is responsible for making the facility available to investigators and agents without their having to attain more than a minimum level of competence in computer system expertise. It is believed that the existence of this group, which is due to the fact that MISS is an information system designed for a particular application and community of comotivated users, will largely mitigate the problems that arise when an ordinary isolated person tries to use a network in its early stages of development.

References

1. Peter Lykos, "Hierarchical Computing: A Computer System Cost/Effectively Optimized for Education," *Computing in Higher Education 1971;* Proceedings of the EDUCOM 1971 Fall Council Meeting and Conference, EDUCOM, Princeton, N.J., pp. 11–14.

2. William H. Bossert, "Hierarchical Computing," *Networks for Higher Education;* Proceedings of the EDUCOM Spring 1972 Conference, EDUCOM, Princeton, N.J., pp. 88–89.

3. Ruth Davis, "Practicalities of Network Use," *Networks for Higher Education,* Proceedings of the EDUCOM Spring 1972 Conference, EDUCOM, Princeton, N.J., pp. 13–28.

4. Herbert A. Simon, *The Sciences of the Artificial,* MIT Press, Cambridge, Mass., 1969.

5. Robert L. Ashenhurst, ed., "Curriculum Recommendations for Graduate Professional Programs in Information Systems: A Report of the ACM Curriculum Committee on Computer Education for Management," *Communications of the ACM,* Vol. 15, No. 5, May 1972, pp. 363–398.

6. Robert L. Ashenhurst, "The Minicomputer Interfacing Support System Project," *ICR Quarterly Report,* No. 33, The University of Chicago, May 1, 1972.

10

DENNIS W. FIFE
Institute for Computer Sciences
and Technology
National Bureau of Standards

Primary Issues in User Needs

The Institute for Computer Sciences and Technology of the National Bureau of Standards is assisting the National Science Foundation* in technical planning to delineate major approaches and to derive guidelines for the trials that may lead to a National Science Computer Network. NBS's primary efforts focus on network management requirements and application criteria for communications technology. User characteristics are being briefly studied for perspective on the objectives for the network and on the types of service it may provide. This will give direction to the network management and communications investigations and will also motivate a closely related study of resource-sharing protocols. NBS will also pursue comparative analyses to reduce the uncertainties involved in the potential spectrum of network experiments and thus to identify the major cost factors and areas of development that should be addressed in the trial network. The goal of the NBS project is to describe by mid-1973 the basic types of experiments that will address the central issues, together with appropriate evaluation criteria, so that the NSF can use this technical framework in selecting an approach to the trial network.

There is enough clear enthusiasm for networking research now to identify a mélange of worthwhile experiments to extend networks in academic computing. But NSF goals[1] emphasize resource-sharing arrangements of broad scope rather than individual technological developments that might remain generally unavailable to the academic community. So the experiments for the trial network should be patterned to produce complementary results and momentum toward an evolving, lasting structure for a national network. How should this aim be related to the needs of users? This is a primary issue. The institutional commitments essential to a national network, which include significant changes in traditional philosophies of computing service, warrant specific analyses of the user needs that could be met.

Several approaches could be utilized to develop information on user needs. Given a comparable operational system, direct observations or measurements of user behavior and patterns of resource utilization would be useful to determine trends and to extrapolate future needs.

* Work supported by the National Science Foundation under Grant AG-350. Contribution of NBS, not subject to copyright.

Of course, usage levels on existing networks now are low, and the statistical significance of observation is questionable. NBS research that complements the effort on NSF network planning includes performance measurement techniques for teleprocessing systems. In particular, NBS is developing a dialogue monitor facility to record and analyze the behavior patterns of users of ARPANET or commercial timesharing.[2]

Direct user requests are another means to determine what needs a network could serve. Such requests might normally lead to experimental services and in some respects this might be considered a purpose of the trial National Science Network.

For technical planning of computer networks, the primary means to project user needs must be a consensus of authorities in the field or a form of market analysis by which future applications or services are postulated and their usage is estimated. The considered opinions of authorities at the EDUCOM seminars will be useful. NBS has made a number of contacts and visits in the academic computing world to exploit the same technique. But NBS is also developing a form of market analysis that focuses on existing, prominent sources of computing capabilities in the academic field. The result will be a descriptive matrix of near-future network services relevant to the National Science Network. Near-future services rather than speculative applications are emphasized from the conviction that the trial network must evolve from existing centers of computer expertise. The matrix will combine available specific and statistical data with state-of-the-art assessments, thereby addressing usage factors and market considerations for various types of service.

Chart 10.1 presents a preliminary list of the potential types of service being considered. Nine major categories have been defined under which a number of specific services fall. Administrative computing services supporting university business management are included for completeness but will not be investigated in detail. Instructional services address usage of computers for classroom support, during which the user provides data entries but does no on-line programming. Program development and testing concern services aimed at the creation, debugging, and validation of computer programs prior to their operational use. Computing services are considered to cover operational uses of completed programs, especially on a recurrent basis in research. Library and literature services address needs to search out textual information, particularly on bibliographic information files. The category of data services covers retrieval from established major numerical or statistical

Chart 10.1. Preliminary List of Potential Network Services

Administrative

Instructional
Computer-aided instruction (elementary/secondary/vocational)
Computer-aided instruction (college/professional)

Program Development and Testing
Basic instruction
Advanced instruction
Disciplinary applications
Network resource development
CAI program support
Cross-computer support

Computing
General
Large-scale

Library and Literature
Library catalog retrieval
Current awareness bibliographic search
Retrospective bibliographic search
Research inquiry referral
Network resource information

Data
Statistical and survey data search
Standard reference data search

File
General data management
Large-scale data management
Text file management
Text analysis

Communications
Teleconferencing
Networks interface conversion

Control and Data Collection
Laboratory automation
Network performance measurement
Resource utilization analysis

---

data bases such as those of the census. File services covers numerical or textual data maintained as a form of personal service to individual researchers. Communication services include teleconferencing and also information conversions at the interface between the NSF network and other computer communications networks. Control and data collection as a category covers any real-time applications in experimental control or laboratory automation as well as operational control of the NSF computer network.

Table 10.1 shows the tentative form for the descriptive matrix. Each type of service from Chart 10.1 would appear, with additional information on the intended usage of the service in one or more prototype resource centers. For example, the intended discipline or the educational level of the intended user would be shown, and also the type of communication access—interactive, for example, or remote job entry, remote batch, or remote computer access. The latter covers "automaton" terminals that serve a number of individual user terminals and therefore may require special capabilities to connect resource centers and users, for example in scheduling and in diagnostics interpretation.

It is expected that a resource center may appear in the matrix under several service categories reflecting the breadth of its capabilities. Information on status and availability will highlight existing network experience and the scale of service that a center is presently estimated to provide.

In assessing the scope of the market for each type of service, the intent is to use rather simple criteria. These may include the number of scientists in a discipline or group of disciplines, from which an estimate can be made of the size of each segment that could eventually benefit from network service. This is useful for signifying relative impact but is not intended to measure economic viability or to judge the number of centers needed to satisfy the market. In addition, a qualitative assessment will be developed of the major factors affecting the near-future growth potential for each service in a national network environment.

The NBS study will use the matrix as concise background for considering alternative groups of services. Each grouping would be selected to represent a potential goal for the resource center development undertaken in the trial network. For example, the network experiments could be tailored to promote one or more of the following:

1. Specific discipline or problem-oriented centers; for example, for chemistry or economics.
2. Special-purpose computing facilities of high capacity; for example, an

Table 10.1. Matrix of Potential Network Services

Type of Service	Intended Usage	Market Scope and Growth Potential	Access Mode	Prototype Sources	Status and Availability
Computer-aided instruction	Elementary, secondary, and vocational	118,000 schools and 51,000,000 students. Cost, service availability, and availability of quality materials primarily limit growth potential.	Interactive	Stanford University IMSSS PDP-10	Operational to east and west coasts, 25 schools and about 300 terminals
				Mitre TICCIT McLean, Va. Nova 800	Large-scale trials being developed for 1973–74.
Large-scale computing	Scientific research	About 300 organizations, 66,000 academic users, 225,000 federal and related industry users.* Service reliability, user assistance, and applications support limit increased utilization of large remote centers.	Interactive or RJE	UCLA Campus Net IBM 360/91	Operational on ARPANET
				UCLA Health Services IBM 360/91	Operational with dial-up access
				Johns Hopkins University IBM 360/91	Operational with dial-up access
Retrospective bibliographic search	General research	About 300 organizations, 66,000 academic users, 225,000 federal and related industry users.* Cost is primary limit to wider use; service availability may limit interactive usage also.	Interactive RJE	LEADERMART Lehigh University, CDC 6400	Operational with dial-up access, several major data bases
				University of Georgia IBM 360/65	Operational on regional network, many data bases

* Based on "National Patterns of R&D Resources," NSF Report 70-46, December 1970, and "Research and Development in Industry-1969," NSF Report 70-18, January 1970. The table is only an illustration of the form.

associative processor or a highly parallel machine such as ILLIAC IV.

3. General-purpose processing functions of high quality; for example, statistical data analysis or bibliographic file searching.

4. Comprehensive remote access to any available computing capability or individual research resource.

These illustrative goals involve distinct benefits in extended computing services that a national network could provide. Similarly, they imply different priorities among communications arrangements and various types of resource development. The fourth alternative, for example, will require very low cost, geographically widespread communications facilities, whereas the first two could be sufficiently economical through the public telephone system. To aid the evolution of development guidelines in the trial network, NBS intends to identify and rank such goals in regard to early attainability, importance to user needs, and economic viability of the services involved. In the process, basic statistical factors needed for communications planning should become evident.

Besides initiating a nationwide distribution of in-demand computing services, the trial network will be concerned with support to meet the continuing needs of users for documentation, problem assistance, and so on. In the past, customer support in academic computing has been provided by graduate assistants in the computing center, occasionally aided by vendors' sales staffs. The function is often loosely managed, and few have developed a smoothly working system to assist diverse users. A national computer network with multiple sources and greater distance between the end user and the system developer (or maintainer) clearly will require new systems for customer support that go beyond the bounds of one institution. Market development is also necessary because the market for computer network services is largely latent and should be stimulated to grow during the trial experiments to yield a tangible measure of successful service. In addition, pooling of personnel and capital and distribution of responsibilities for resource development among institutions will give impetus toward new management forms.

NBS is developing alternative organizational models that would effectively implement these functions within the academic and government community, giving some consideration also to the potential role of commercial companies. Of course, the organizations that will actually govern and operate the National Science Computer Network will evolve from traditional interuniversity and government working relationships, undoubtedly under significant political and legal influences. Nevertheless, planning is needed for management guidelines that could accelerate the conclusion of organizational agreements and that will give direction to research and development on supporting technical facilities such as retrieval and dissemination of resource information, monitoring of resource quality, testing of software, and the like.

In summary, NBS is addressing user needs in terms of potential network services and accompanying organizational arrangements to provide user support and essential marketing. The identification and evaluation of the more promising alternatives will clarify the role of a national network in the near future and will assure viable experiments directed at the more difficult issues. At present, the effort has produced an extensive bibliography.[3] Preparatory technological surveys of communications and network management will be completed early in 1973, and representative experiments will be delineated by fall 1973.

References

1. D. Don Aufenkamp and E. C. Weiss, "NSF Activities Related to a National Science Computer Network," *Proceedings of the First International Conference on Computer Communications*, Washington, D.C., October 1972, pp. 226–232.

2. M. D. Abrams, George E. Lindamood, and Thomas N. Pyke, Jr., "Measuring and Modelling Man-Computer Interaction," First Annual ACM/SICME Symposium on Measurement and Evaluation, Palo Alto, Calif., February 1973.

3. Robert P. Blanc, Ira W. Cotton, Thomas N. Pyke, Jr., and Shirley W. Watkins, *Annotated Bibliography of the Literature on Resource Sharing Computer Networks*, National Bureau of Standards Special Publication, Washington, D.C., September 1973.

11

WALTER C. HAMILTON*
Brookhaven National Laboratory

Large-Scale
Numerical Analysis
as Applied to
the Basic Sciences

In the basic sciences there are many problems that require extensive digital computation but little input and output. Such problems in physics, chemistry, biology, applied mathematics, and engineering are "compute-bound" rather than "input-output bound." In some of these problems, at intermediate stages in the calculations many numbers must be stored temporarily and accessed very rapidly.

Large-scale, compute-bound problems are remarkably similar in their broad characteristics from field to field. Most of them can be classified into one of three categories:

1. Multidimensional integration of differential equations and numerical evaluation of many multidimensional integrals.
2. Manipulation of large-scale linear systems as in least-squares and linear programming calculations.
3. Pattern recognition in complex systems.

Although these computational styles are used in each of the basic sciences, multidimensional integration is most characteristic in mathematical physics and theoretical chemistry; manipulation of large-scale linear systems in crystallography; and pattern recognition in physics and biology. Two requirements for most of these problems are a high-speed central processing unit and large amounts of fast memory for the storage of intermediate results.

In my own field of interest, molecular structure, specific illustrations of the use of multidimensional integration can easily be drawn. In one extreme case, the molecular quantum chemist wishes to calculate from first principles the electronic wave function of a complex molecule as a function of the coordinates of the nuclei. The input for each nuclear configuration is the x, y, z coordinate of each atom in the molecule. The output is the energy of the molecule as determined by the integration of the Schroedinger equation $E = \int \Psi H \Psi$. The usual technique is to expand the molecular wave function in a set of atomic functions:

---

* Died, January 23, 1973. The transcript of Dr. Hamilton's talk was edited for publication by Peter G. Lykos of the National Science Foundation.

$$\Psi = \sum_{i=1}^{n} C_i \phi_i,$$

where Ψ is the molecular wave function, C_i is a linear coefficient, and ϕ is an atomic function of the x, y, z coordinates of an electron. In order to determine the optimal values of the linear coefficients C_i, first it is necessary to evaluate six-dimensional definite integrals like the following:

$$\int \phi_i \phi_j \, H' \phi_k \phi_l d\tau,$$

where ϕ is an atomic function of the electron's x, y, z coordinates and H' is an operator component of H. Determination of the electronic wave function of a small organic molecule may require several hours of computational time on even the largest computer system available today.

An example of the second type of compute-bound calculation, which also requires large input capability, may be drawn from the study of protein crystallography. Tens of thousands of data points may be measured by x-ray diffraction of one crystal sample. One must calculate a three-dimensional Fourier series using the 10,000 data points as coefficients (which yields 1 million points defining an electron density function) in order to make a rough determination of the crystal structure. This rough structure may then be refined, using iterative least-squares techniques, to determine perhaps 2,000 parameters from the 10,000 data points. The required calculation of an inversion of a 2,000 by 2,000 matrix can consume tens of hours on a CDC 6600 computer. Again the amounts of input and output are relatively small.

The third kind of problem is illustrated by a specific example drawn from high-energy physics. The scientist uses an accelerator to produce a particle which traverses a bubble chamber and produces tracks. Photographs are taken of these particle tracks at very frequent intervals (perhaps 10,000 photographs for one experiment), and the information is digitized and stored. The computer is then used to identify specific events that have taken place as indicated by particular complicated particle tracks. Even with fairly coarse resolution, a 1,000 by 1,000 grid (or 1 million different points) may be required to represent each encoded photograph. Although researchers initially used the computer alone to try to recognize and analyze these complex patterns, more recently they have been studying computer-generated displays of the photographs to pick out patterns before submitting the photographs to the computer for analysis.

Interactive display units connected to a computer for modeling molecular structures have proved quite useful at Brookhaven National Laboratory. Research scientists have been able to rotate images of molecules on the screen and view the effects of various configurations of molecular structure on potential chemical activities. Interactive display units have also been very useful in presenting the results of the quantum mechanical calculations described earlier. The display unit has enabled scientists at Brookhaven to see immediately the results of changing a few parameters on a potential energy surface that governs molecular rearrangement or a chemical reaction.

I believe that networking can provide important resources for users with compute-bound problems. Why? Most university computer centers simply do not have the equipment to do the job. Indeed, some very good scientists are being constrained by having computer power well below what is possible from existing technology. Convenient access to the largest computers available would be a tremendous asset to these people. Since many of their computations involve extremely sophisticated algorithms, access to standardized, debugged, well-documented programming systems also would be a big asset. Maintenance of large data bases at a few centers and convenient access to these data bases would greatly stimulate the applicability and efficiency of science in some fields. Such uses will probably increase greatly over the next several years. Although most detailed graphical manipulation may be done on local computers, access to "number crunchers" and large data files will also be of great importance.

Will networking expand our scientific horizons to new classes of problems much as the computer did in the first instance? Possibly so, but probably not so much through the network concept itself as through the power of new types of computers that networking will make available to many users.

12

Computer Usage in the Natural Sciences

Report of Workshop 1

JULIUS ARONOFSKY
Southern Methodist University
Faculty Discussion Leader

Contributors:

WINIFRED ASPREY
Vassar College

RONALD BLUM
Duke University

J. A. DISBROW
National Oceanic and
Atmospheric Administration

WALTER C. HAMILTON
Brookhaven National Laboratory
(died January 1973)

FRANK HARRIS
University of Utah
Salt Lake City

ROBERT E. KAHN
Advanced Research Projects
Agency

PETER G. LYKOS
National Science Foundation

JAMES N. SNYDER
University of Illinois,
Urbana

JAMES W. TELFORD
University of Nevada,
Reno

Introduction

The use of computers in the area of arithmetic or symbol-handling problems in the natural sciences covers a broad spectrum of users. In addition to the individuals who require large blocks of computer time or memory, there are many small users who in the aggregate require extensive computer power. We feel that a properly developed network can accommodate both classes of users and that neither should be neglected in characterizing computer needs.

From the standpoint of a network able to distribute the necessary capability from many computer centers, the data transmission aspects are of prime importance. Typically, problems that utilize large computer systems frequently have small input and output requirements. Sometimes the large system users need file systems with continuing input and output for updating and integration procedures. Systems involving many small users may not make such heavy individual demands on the computer facility but will require rapid communication and resource allocation to permit fast turnaround and interactive use.

The goal of a network should be to offer access to the best method available for a broad class of diverse users.

Massive, Large-Job Computing

The participants in Workshop 1 identified some of the types of users who require access to large-scale computing systems as follows:

1. *Theoretical chemistry*. Several classes of large problems involve extensive calculations. Sophisticated problems taking full advantage of hardware capabilities are required. These problems include ab initio calculations of the electronic structure of molecules and Monte Carlo calculations in statistical mechanics and reaction rate problems. Trivial amounts of input and output are usually required.

2. *Crystallography*. Calculations involve the evaluation of three-dimensional Fourier series at millions of data points and the use of thousands of coefficients. The computations are massive, and the use of graphics technology is important.

3. *Experimental physics*. Very large experimental data bases are often available from cosmic ray or high-energy physics experiments. The abstraction and parametrization of interesting events from the large number of data make great demands on both computer speed and memory size.

4. *Geological and space sciences*. Again, very large data bases may be

available and are used for the formulation and testing of models. These tasks involve the integration of complex differential equations. In some areas of research the methods for data analysis are still ill-defined and man-machine interaction may prove to be of great importance.

Remote sensing of the earth's surface and atmosphere is providing enormous quantities of raw data that will be used in the manner described earlier.

5. *Biology.* Aside from large problems in statistical analysis, theoretical work in biology may include computer modeling in areas such as physical chemistry of intracellular processes, membrane transport, and the prediction of the conformation of macromolecules.

6. *Computer science and applied mathematics.* Mathematical methods, applicable in all natural sciences, the development of higher-level languages, and methods of symbol and equation manipulation can utilize large amounts of computer time.

7. *Engineering.* Computer usages in engineering include the analysis of given systems and the synthesis and design of new systems based on known empirical data. General requirements are high computer speed and large amounts of storage for intermediate results. In some applications large input data bases are of prime importance, and in several areas interactive graphics are becoming essential.

Many Small Computing Jobs

The use of computers in classroom and other educational activities is qualitatively different from their use in research-oriented contexts, but quantitatively they can become as large or larger in the aggregate than the research applications.

Any student in postsecondary education in the United States should have the opportunity to achive computer literacy. Faculty and students should appreciate the use of the computer as a tool, the value of computer sciences as an academic discipline, and the significance of both for their chosen fields.

We feel that given appropriate facilities and encouragement, 20 to 50 percent of the postsecondary population can achieve this level of computer literacy within the next decade. We base this feeling upon the experiences at such schools as Vassar, the University of Illinois (Urbana), and Dartmouth.

To achieve this objective there are several avenues open: local computer centers, regional or statewide computer networks, and national computer networks. We recognize the following classes of "small users": (1) university students who are using computers in (a) courses on the introductory level, (b) courses in computer sciences, or (c) advanced subjects involving research; (2) small colleges with limited individual campus facilities; (3) postsecondary institutions with no computer facilities.

An important question is what are the advantages of regional or national networks for these different classes of users. For students in classes 1a and 1b there might be flexibility and economy in networking. For students in class 1b, moreover, national networks would be very desirable in order to take advantage of the vast range of resources that could be made available on a national scale. Class 1c could probably make good use of a vast spectrum of resources with the proper organization of a network.

In class 2 we frequently find faculty who are involved in research and who desperately need the kind of computing power that could be made available on some kind of network, although not necessarily a national one.

For faculty in class 3, a vast educational effort to indoctrinate and enlighten them is needed. This may best be provided on a regional basis and may mean supplying computer services through a regional network.

Diversity in Computing

Besides those of the groups just discussed, there is a broad spectrum of needs and a wide variety of techniques for meeting them.

An important class of users consists of students doing research projects. The word "student" is broadly interpreted to include the more innovative undergraduate students, graduate students (particularly those in the early stages of their graduate career or enrolled in professional degree programs), and people involved in continuing education or advanced training sessions. By research projects is meant situations where the students transform data for interpretation, creation, and testing of models, actively participate in a game, or use programs and associated data bases generated by researchers working at the frontiers of knowledge.

This class of computer users has requirements broad enough to encompass most kinds of computer resources. The class encompasses computer science research groups as well as a high school group involved in a special project in the physical, biological, or social sciences.

Access to programs and data bases of researchers, particularly of those working in the problem-solving and decision-making disciplines, constitutes networking's most exciting prospect. Networking could make possible a more direct link between the ferment, excitement, and enthusiasm of research and the creation of knowledge, on the one hand, and the student or the user of new knowledge and the techniques whereby it is gained, on the other.

In addition to access to special-purpose languages and utility and application packages that have rather specific and localized system requirements, a major use will be simulation and modeling involving data bases. This is being done now primarily in the social sciences but will be done more and more in the natural sciences. An extension is computer gaming or training exercises that involve more than one participant or source of data at different geographical locations. Ultimately this trend leads to computer-based conferencing as an aid to problem solving and decision making, making possible smoother-working, discipline-oriented communities of scholars.

Networking not only provides the means whereby improved communication and resource sharing can take place, it also requires the discipline of standardization, and it introduces the possibility of redundancy and hence improved reliability beyond that provided by a local stand-alone facility.

Two major problems that loom very large are (1) the probable fragility of programs that are part of a research experience and (2) the probable lack of documentation that would hamstring the network information center necessary for a network to function.

Some workshop participants felt that universities are more likely to join networks than colleges are. Communication costs are quite variable and unfortunately are most serious for those institutions that could benefit the most. Research needs for computing support are most likely the first for which networks will be justified.

A second observation was that there does not seem to be general awareness of access possibilities to the computing resources of national science laboratories. The National Institutes of Health support the BIOMED activity at the University of California at Los Angeles with an IBM 360/91; the Atomic Energy Commission is supporting Lawrence Radiation Laboratory with a CDC 7600 that is remotely accessible and has been declared a national computing facility; the National Science Foundation is supporting the National Center for Atmospheric Research with a computer available to university researchers; the National Aeronautics and Space Administration is supporting the Langley Research Center at Hampton, Virginia, whose Universities Space

Research Association has formed the Institute for Computer Applications in Science and Engineering; the Advanced Research Projects Agency is supporting the ARPANET; and the list goes on. A catalog is needed of what computing resources exist in national science laboratories and under what circumstances they are available to higher education.

13

WILFRID J. DIXON*
University of California
at Los Angeles

Data and Computing Facilities

Introduction

Whatever the field of application, statistical problems, modeling, and retrieval, all require certain basic data facilities. The subject matter may be physiological systems, therapy, sociology, health service systems, clinical trials, or psychology, but the basic processing requirements are much the same. The basic functions include data origination, entry to file, editing and screening, reorganization, preanalysis, analysis, display and reporting, storage, and data cleansing and reduction. These functions form a good framework for discussing the needs and special resources required for data banks, statistical analysis, and behavioral and clinical modeling.

The potential advantage of a network is that the user would have access to the best facilities for his individual job and that the computer center would be free to specialize and thus to provide selected services more effectively. Centers may serve users' needs adequately but specialize in only a few of the functions listed and will almost certainly specialize in only a few fields of application. The system should, of course, become better designed for the specialized tasks. It is difficult to see, however, how specialization could be extreme in the main areas of this workshop: data banks, statistical (or data) analysis, and modeling.

The user will require not only certain hardware capacities but also systems software that will allow him easy access to the hardware. Furthermore, he will need generalized applied software systems that will minimize his needs for his own applied software. Beyond this, he will need access to specialists who understand his field of application and who have experience in meeting the needs of particular classes of users. But perhaps most important, the user will need information on the resources available to him, the location of experts who can consult with him on handling his problems and can provide help in reaching these resources.

An applied program may be "written" in three days, but it may be a year before that program is generally useful. The same is true of higher-level computer languages and retrieval systems. The same effect probably holds for networks. For example, even though the Advanced Research Projects Agency network is operative and is used, it solves

* I wish to acknowledge the extensive assistance of Dr. Patricia Britt in preparing this paper.

only part of the problem in developing a generally useful network system.

Some Examples of Network Demands

A user may wish to gain access to a particular facility for any one of several reasons. His research projects may be complementary to those of the facility, and he may be interested in a cooperative venture. He may want to use the facility's applications programs and consulting services. He may need a specific service offered by the facility. Or he may require help in setting up, interpreting, and selecting the appropriate analysis for programs exported by the facility. This kind of service could involve an intensive start-up effort, a continuing occasional interactive consultation, or both.

Network demands have been generated by a number of the users of the Health Sciences Computing Facility at the University of California, Los Angeles. Some examples may illustrate the requirements:

Tissue Typing

The use of computer analysis for the classification of sera and typing of subjects for transplantation antigens provides a basis for several studies. For example, Dr. Paul Terasaki's studies supplied the first objective evidence that the lymphocyte antigens act as transplantation antigens in kidney transplants. To carry out these studies, it is necessary to type prospective donors and recipients prior to transplantation, and to this end samples are received from many transplant centers throughout the country. In furthering the research aims of the project, a by-product is the furtherance of medical care by influencing the selection of donors.

A particularly important problem is whether adequate matches of cadaver organs can reasonably be expected. Consequently, the typing of potential cadaver donors is of great research interest, and a substantial effort is devoted to aspects of this problem. In the typing of such donors it is necessary quickly to select the best-matching recipient. The larger the recipient pool, the more likely it is that a suitable match will be found. As the size of the pool increases, the search task becomes more demanding. Special files are maintained to facilitate this search. After the typing, results are available for the prospective donor, and the file can be searched for suitable matched recipients in a matter of seconds or minutes. As the recipient pool increases in size, the importance of this ability to search rapidly will also increase.

The analyses upon which the tissue typing are based require extensive support in hardware capacity, analytic programs, and staff expertise.

The typing service, now offered to a rapidly increasing number of centers, requires this base as well as the file-searching capacities.

Therapy Optimization

There are well-known areas of radiological or medical management where the course between efficacy and toxicity may be difficult to steer. Many factors complicate the improving or optimizing of therapy in such cases, and measurements of response are difficult to define or obtain.

Because of these difficulties, attempts to cope with such problems tend to be more exploratory than definitive. Their payoffs cannot be guaranteed. Yet the possible benefit to patient care if even a small percentage of these studies bears fruit is such that one is reluctant to eliminate investigations of this type. One of the advantages a large computer facility has over a number of smaller project-dedicated machines is its capability for supporting exploratory investigations at times when the computing load permits and with staff that finds in this a challenging supplement to its regular activities.

One such project undertaken at the Health Sciences Computing Facility (HSCF) relates to improving dose distribution from radium implants. Radiologists frequently attempt to improve dose distribution to various parts of the tumor by removing some implants earlier than the remainder. Even well-planned arrays of implants fall short of what is desired because the toughness and elasticity of tissue make it difficult to place the implants precisely.

Implants can be partitioned into two or three groups, and the time for removing each group is calculated to minimize the undertreated and overtreated areas within the tumor volume. The program has been completed satisfactorily for the case where the radiologist selects needles for early removal and where a special effort need not be made to spare a specific nearby area of tissue. The investigation of algorithms to automate or assist the identification of implants to be removed has been greatly facilitated by interactive graphics support.

A treatment-planning program based on this research is now available at HSCF to clinicians with IMLAC graphics terminals. It has been under intensive test for some time and will soon be suitable for general use.

Interactive Exploration of Diagnostic Problems

There is at present no reliable way to predict the outcome of the various methods of treatment for schizophrenia, and accurate information on their comparative costs is not available. A treatment that is cheap to give may actually cost the most in the end because the patient has to stay in the hospital longer.

Improved methods of prediction of (for example) length of hospital stay, probability of release, and cost for the different treatments and for different types of schizophrenia would be of great practical help in administrative and budgetary planning and to the clinician in selecting the right treatment for a patient. It would also be of great theoretical value in throwing light on some of the important dimensions of this malignant illness.

This problem has been studied extensively by investigators using HSCF. It is typical of a class of problems needing solution. Many of the computer needs of such studies can be met effectively only at a facility that specializes in providing the services—hardware, software, applications programs, and consulting—required for facilitating data handling and analysis.

Research Activities at The City of Hope

To list the programs that have been written or are being developed at The City of Hope would scarcely convey either the scope of activity or the impact the terminal is having on the institution. Computing equipment, if intelligently introduced, brings not only efficiency and cost savings but, subtly, a change in thought. That is, one tends to structure activities to make use of the resources at hand. At first the investigator does the obvious. Then the simple statistical results act as a catalyst to change his aspirations, and he seeks more from his results and equipment. Then comes the reflection that alters the methods and procedures, and finally, in its wake, the innovation wherein new concepts and ideas not before considered are seen as feasible. Such stages are common with an able and inventive group, but in an establishment that is strong in ability yet modest in budget the lack of a large machine would exclude that type of growth. A terminal with access to a large system can provide an excellent compromise.

Brief summaries of some of the main research problems being undertaken via the terminal follow:

Statistics have been used to analyze the genetic experiments performed by irradiating drosophila to produce dominant lethals, and in a number of related experiments in mutagenics.

The reduction of arithmetic efforts in various analyses to that of punching cards corresponding to readings (and the eventual automatic punching of these data) provides more time for analyzing the reasons for such calculations. Such is the case with *lipid content* in a variety of tissues under various conditions, both normal and diseased, and many other biochemicals.

Gene frequencies for galactosemia are under analysis where it is necessary to look at semiquantitative screening data, assay data, and electrophoretic results to determine frequencies of the various ranges of enzyme activities. This and other analyses could not have been undertaken without the terminal.

In the realm of *cardiology,* some *hemodynamic questions* concerning the stress relations in blood vessels and other mechanical restraints and entities are under active experimental investigation. Theoretical tools are needed to further the effort.

Of considerable import are questions that deal directly with clinical treatment. Among these are the researchers in *respiratory physiology* centering around a lung model, using clinical data collected in the inhalation laboratories.

Much effort is devoted to treatment techniques in patients undergoing *radiation therapy.* Of particular interest are cases where the beam passes through bone, lung, and various air gaps. Unique in this is the calculation of the ionization distribution at the level of the Haversian canals in bone.

Complex technical problems such as reading the spots obtained from *electrophoresis* and similar physical chemical techniques are being tackled with an eye to using the computer as the pattern recognizer and strategist. If this research proves successful, a minor revolution will immediately change this field.

In addition, biological organization presents an enigma that may well begin to challenge the resources of the machine. Simple strategies are contemplated that taken together form a type of group behavior such that more complex strategies can be generated. These logistic patterns are now no longer simple, and this richness of structure admits breadth of application and an elevation above the obvious. Here, then, is a problem that needs a large computer. In attempting to generate such a scheme, the staff at The City of Hope considers itself a part of the general campaign to reduce the unknown to a theory of biology that will set the stage for manipulation, which is the art of medicine.

The New Statistics

The major specialization at HSCF continues to be the development of statistical tools. Significant advances in statistical methods have resulted largely from the availability of new and more powerful computational tools and from new challenges arising from attempts to apply accepted methods to specific research problems. To an extent perhaps greater than we as statisticians care to admit, the statistical problems we have attacked, the methods we have used to attack them, and even the way we have thought about them have been affected by these factors. Thus, in the 1930s the desk calculator provided an adequate tool for balanced analysis of variance designs and small regression analyses, which were applied, with admirable ingenuity, to problems being raised in the fields of agriculture and psychology. Later the card sorter stimulated the development of contingency table analysis, especially applicable to research problems in medicine and the social sciences. Meanwhile, research problems in biology provided the stimulus and direction for a number of topics in mathematical statistics such as multivariate analysis and decision theory.

But the resulting methods remained largely untested. Theoretical problems were frequently proposed and solved, but only the most modest practical applications were possible because of the lack of computing power. Not only did this leave the biologist without assistance on real-sized problems, but research in some areas of statistics also stagnated since the techniques could not be put to a test. The statistician often thought that the problem would be solved if and when the computations became feasible. Now in many cases adequate computing support has instead demonstrated the limitations of the approach itself. Solution of the problem requires an alternate technique.

Thus, the introduction of the modern computer into the biological environment put the familiar techniques of the statistician to a severe test that frequently found them wanting. Many problems were a lot bigger than expected. Frequently the basic problem was not how to look at a small set of data optimally but how to look at a large set of data reasonably, or at all. Missing data became a fact of life, not a random accident. Bad data caused even greater problems. The thin tails of the cherished normal distribution grew thick with real data outliers. Discrete and counted data were surprisingly common and were frequently mixed with continuous data. And even the continuous data often required transformation to make them acceptable for analysis.

The difficulty of analyzing biological data resides not only in their complexity, magnitude, unavoidable and often conditional incompleteness, and large measurement error. The basic interactions of variables also often depart markedly from those assumed by classical statistical techniques. Indeed, much of the effort of biological research is addressed to characterizing these interactions. They often cannot be phrased explicitly, being couched in systems of differential equations or in simulation constructs. A vast new research area is opened by this need to develop new statistical and modeling techniques to guide efficient experimental planning and the development of validation criteria for testing of realistic models.

The introduction of the computer immediately challenged the statistician with a host of unfamiliar problems and with masses of alien data. It also presented him with a powerful tool. The designation, the new statistics, represents the statistician's response to the challenge and opportunity presented by the computer. The response has only begun, and the work at this facility is a significant fraction of the response.

A number of innovative uses of the computer in the solution of statistical problems are under way at HSCF. These developments include

• Linear scores from preference pairs: The introduction of judgment into the analytical process.

• Typical values: An effective compromise between the fully specified parametric model and the non-parametric model, preserving the advantages of both.

• Boolean factor analysis: Factor analysis of dichotomous variables, subject to error. Individuals observed (with error) on a number of dichotomies are classified into groups by their patterns of attributes.

• Clustering: Definition of groups by similarities in variables or cases or both.

• Stepwise procedures: Sequential selection or omission of variables and cases for optimum fit of linear and nonlinear models.

• Winsorizing: Methods for analyzing data containing errors beyond the basic statistical error model.

• Least p^{nth} power regression: Least-squares analysis is not optimum in many biological situations. Adjustment of p to the basic measurement distribution can provide improvements.

• Missing data: Techniques for adapting incomplete data vectors to standard statistical techniques.

• Permutation procedures: Generation of estimated sampling distributions for new estimates and procedures.

A network would make such tools available to the wider research community much earlier, speeding both their fuller development and the research that depends upon them.

A network approach is made all the more necessary as well as natural by the special demands and promises of interactive computing. Until quite recently, the advances in computing systems most significantly affecting their use in research were a steady increase in computing speed, the development of large disc files for program and data storage, and the gradual evolution of programming systems. All of these provide greater flexibility and convenience to the user. These developments continue, but much more slowly, while the most significant current trend appears to be the development of an "interactive" computing capacity.

The strategy of interactive computing differs significantly from that of batch computing, the mode of operation that was almost universal until quite recently. When a user submits a batch job, he attempts to construct his programs to provide for all contingencies; he submits both program and data to the computer, and awaits his completed results. If his problem requires that he exercise judgment between steps of the computation, it may require many iterations to complete. The user of an interactive program can communicate with it from his terminal while it is being executed. He can monitor the computation, specify options, and provide additional data as required during the process, thus performing a sequence of computations in one step. Properly designed, programs operating in this manner can permit the investigator to pursue a variety of hypotheses in working toward a solution. Indeed, his direct involvement with the computation often suggests alternate hypotheses he would not otherwise have considered.

Interactive computing depends on a number of system features: the availability of a satisfactory terminal and of sufficient storage in the system for programs and data, the ability to time-share the most expensive resources of the computing system, sufficient computing power to provide reasonably rapid response, and general-purpose conversational programs to handle file manipulations, user communication with the system, and common computational tasks. But the most important requisite is a fresh approach to program structure and logic. The methods that were most fruitful in designing batch programs produce interactive programs that, while they may offer the attractions of novelty and faster service, fall far short of the obvious promise of interactive methods. Many early attempts at interactive computing were either batch programs that talked to the user from time to time or somewhat extended desk calculators available at a computer terminal. Although both these facilities may be quite useful, neither really constitutes a significant advance. Nonetheless, interactive computing can be a dramatically successful research tool, drastically reducing the time span from problem formulation to solution, making possible the solution of problems for which analytic techniques are not yet adequate, and promoting discovery through providing the user with greater insight into his data.

During the past few years we have explored interactive statistical methods with these considerations in mind. The ability to time-share the resources of a powerful computing system among groups of jobs having diverse requirements and including both batch and interactive jobs has been the basis of this work. Without this, the kind of interactive programs we have been developing would not be feasible, since they are too demanding to be handled by a small dedicated system but are too slow-paced (because of the human interaction time) to justify sole use of a computer of sufficient power to handle their peak computation loads. To this end, a general-purpose time-sharing system has been developed. In addition to managing system resources, it provides for communication between user and system, controlling terminal access to the system. Conversational packages for specifying jobs, manipulating data files, and examining program output are provided as part of the basic system, and a file management routine provides the user with a means for creating, updating, and using on-line data files. This system provides a flexible basis for interactive statistical programs, which can then be oriented to the intended applications and need not themselves incorporate system features. We now have several interactive statistical packages that are ready for general use, and we can begin to evaluate our interactive techniques.

Consultation

A network will allow not only facilities but consultants at facilities to specialize more finely and become colleague researchers with leading investigators. For example, at HSCF staff members can specialize in only one area like medicine and radiology, psychiatry, physiology and pharmacology, dentistry and nursing, psychology, public health and sociology, neurosciences, genetics, clinical trials, or administration and health services.

Consultation and collaborative research is the basis for the selection of our facility research. Cooperation between users and HSCF staff members occurs on a number of levels and in several modes, including the following:

• *Technical information and advice.* Many users need advice on experimental design, data management, program submission, and interpretation of output. We supply this support on a limited basis free of charge and on a more extended basis for a fee.

- *Collaborative research.* A deeper involvement with various researchers often develops. This is encouraged for projects that contain challenging new problems and that stimulate new statistical or mathematical research.
- *Console-assisted joint research.* When sufficient interactive analytical techniques are available at a console, it is possible for the researcher and consultant, while together, to complete their analyses, verify various subhypotheses, and so on, by performing immediately any analysis suggested by either.
- *Remote tandem console consultation.* Although not physically together, extensive consulting and collaboration will be possible by providing tandem console operation. This can be a continuing arrangement or a short-term service to assist with a specific problem.

Summary
The major question is how to provide access to specialized computing power most effectively. A number of considerations combine to recommend organization along functional rather than regional lines. The research community requires both software and consulting support. Insofar as it is economically feasible, those types of specific support most needed locally probably should be developed and maintained locally, making use of package programs and other available software resources. But as backup to local expertise or as primary support for other investigators, it makes sense to encourage centers that are committed to research and development in a given software area to provide computer-supported access to that software as well as to the experts who create and maintain it. Distance does not impose any problems other than those of commercial telephone rates. A network could mitigate these problems.

Outside of the administrative problems in setting up such a network, the most important considerations are unquestionably in the area of interface to the users. This includes a wide spectrum of problems: letting people know what is available, what it is good for, and how to use it; providing consulting services; making interaction with the network as simple as possible; and providing demands for feedback. If these problems are ignored or given only passing consideration, a network would be of marginal value at best.

14

Numerical Data Bases, Statistical Analysis, and Modeling

Report of Workshop 2

MARTIN GREENBERGER
The Johns Hopkins University
Faculty Discussion Leader

Contributors:

MURRAY ABORN
National Science Foundation

JOHN C. BERESFORD
DUALabs

WILFRID J. DIXON
University of California,
Los Angeles

MICHAEL A. HALL
Lawrence University

DAVID R. HARRIS
Texas Christian University

JOAN G. HAWORTH
Florida State University

DUANE F. MARBLE
Northwestern University

NORMAN NIELSEN
Stanford University

JAMES A. PETERS
Smithsonian Institution
(died, December 1972)

DOUWE YNTEMA
Massachusetts Institute
of Technology

Synopsis
Users who employ computers for working with numerical data bases and doing statistical analysis and modeling are characterized by a heavy reliance on program packages produced by others and on a need for flexible file management capabilities for updating and reorganizing their data bases. This leads them to place high importance on good program documentation, self-explanatory error messages, guaranteed stability of programs and data bases, ready availability of individual guidance and assistance, and easy access to the programs and data of others. These users do already share resources with one another, but generally by physically moving data or programs, which introduces complications and impediments that increased networking could help to eliminate. Networking could also help to encourage greater intellectual contact and communication among these users irrespective of their physical location.

Such benefits will not come automatically. The current means for making computation available does indeed leave much to be desired; but networking by itself is hardly a panacea. Networking presents many problems: problems of security and accuracy; unexpected effects on the user and on how the computer is used; heightened requirements for good documentation and support; and a host of political, organizational and managerial problems, as detailed in the workshop report. These problems need careful attention. Networking provides a vehicle within which solutions to current deficiencies can be worked out, not a solution itself. Intelligent experimentation and analysis are important. A variety of possible tests and demonstration projects is suggested. The involvement and support of constituent groups is essential.

Introduction
In addition to sharing the use of computer hardware, scientists can and do share data bases, computer programs, and their expertise in the use of these tools. They frequently reach the limits of what their own institution can provide and require the resources of other institutions. Until now, such resource sharing has been accomplished by the physical movement of programs, data, and research manpower. The choice has been between moving data or programs to one's own installation or moving it to another installation that has the hardware or software capability required. An informal network already functions via telephone, airplane, and third-class mail for this type of resource sharing.

Adapting a program of any complexity to another computer is never a trivial task, and moving a large data base is a tedious chore at best.

The experience of EIN (Educational Information Network) indicates that researchers find the process of moving data to another installation and working by telephone or mail with programmers there very frustrating.[1] They prefer to take the time and trouble to move and adapt the programs they need to their own environment. This procedure seems inefficient and costly compared to the possibility of their using programs remotely over a well-managed, low-cost national computer network.

In order to specify the desiderata for a national network from the point of view of behavioral scientists and others who use computers in similar ways, the workshop examined the characteristics and needs of this class of user and drew conclusions about the direction that network experiments should take to further research and help understand key problems. The workshop also considered the constituent groups whose participation and support might be critical to the cause of improved resource sharing through networks.

User Characteristics

The computer user in the behavioral sciences is concerned with the origination of data, its editing and reorganization, entry and storage of data files, display and reporting, preanalysis and analysis, data cleansing and reduction, and dead storage. These functions cut across the activities of statistical analysis, modeling, retrieval, and simulation. Yet few computer installations provide more than one or two of these functions in formats convenient for computer users. Most installations still require users to write their own software, design their own file management systems, and learn the computer-oriented control statements of a statistical package.

A recent study indicated that only 16.9 percent of the faculty and 23.6 percent of the graduate students in social science departments on 130 campuses across the nation were able to program in a high-level language. Only 18.8 percent of the faculty and 41.3 percent of the graduate students could use the statistical packages available at their institutions.[2] It is not surprising that less than half the faculty were reported to be using computers in their research. They do not have, and generally do not wish to have, the programming expertise that (unfortunately) is expected of a user.

Behavioral scientists often work with large numeric data bases with built-in growth factors, whether based on recurrent cross-sectional observations (for example, the U.S. Census) or continuing, high-volume time series (for example, daily stock market prices). Updating and tapping these data bases can pose major problems, and adequate file-management systems are rare even for users who are willing to do their own computer programming and systems design. It is common to use elaborate manual segmentation systems to accommodate large data bases to small address spaces. Virtual memories may help but are not a complete answer.

User Needs
Among the most serious computer problems facing behavioral scientists today are (1) inadequate standards for data documentation and problems of data transfer, (2) insufficient programming knowledge and interest, (3) a slow rate of diffusion of computing innovations, (4) program inaccuracy, poor documentation, and difficulty of transfer, and (5) inadequate software for many types of complex processing operations. In addition, certain applications are seriously constrained by deficiencies in data storage technology and file management systems.[3]

Almost every computer installation provides a standard, easy-to-use statistical analysis package. Of the departments using computers cited in the aforementioned study, 76 percent report having access to the BIOMED package of UCLA. But this and other nationally distributed statistical packages are not interactive, and they do not facilitate preanalysis as well as analysis in an efficient manner. There are some very useful interactive tools that do give scientists flexible control over their procedures and data, but their development is still in an early state.

These problems may be translated into the following list of user needs:

1. Users need systems software that provides easy access to programs and data, adequate error messages to allow personal debugging, standard routines for reformating or recoding data files and translating "stranger" codes, a built-in instructional file of directions on how to use the systems available, and relief from having to worry about the compatibility of different character sets.

2. Users require stability in the files they use, whether programs, data bases, or systems. They should not be required to learn new sets of control statements or have to maintain their own personal copies of older versions of the file they are using.

3. Users need files that have been tested for quality and are well documented. A referee group of peers within a discipline could be asked to evaluate files such as the BIOMED package, SIMSCRIPT, or the 1970 census data to certify their acceptability.

4. Users require files that often cannot be easily relocated or replicated. For example, the Cambridge Project at MIT currently is assembling a

set of behavioral and statistical programs that will be difficult to relocate, especially during the developmental period, when it would be desirable to have it tested by a wide spectrum of potential users. Project IMPRESS at Dartmouth is extremely difficult to remount at other locations. The financial data files at Chase Manhattan Bank and the economic time series data of the National Bureau of Economic Research are further examples.

5. Researchers need interactive systems that allow them to browse through data files, manipulate a system, or change parameters or variables and pose questions on-line in simulation models and statistical analyses.

6. Users require an information center on available resources. This center may or may not be computerized and on-line, but it must contain references and abstracts on programs and data bases, a system access procedure, individualized guidance for special problems, and, possibly access to computer specialists who understand the user's field.

Shareable Resources

A number of large or unique data bases could be rendered more useful if made available over a network. Although they are in machine-readable form, some are not kept on-line on a regular basis. This situation might change if the network stimulated increased demand for them.

The largest collection of such data bases is the information released by government agencies. The annual output of the Bureau of the Census alone is immense. For many researchers, data from the Social Security Administration, the Internal Revenue Service, the National Office of Vital Statistics, the National Center for Health Statistics, and the Departments of Commerce, Labor, and HEW, among others, are vital to the progress of their research. The National Science Foundation has access to data bases collected in connection with research it has funded.

Such data bases are also available outside government. A few of these places are the Interuniversity Consortium for Political Research and the Roper Center for political science and sociological data, the University of Washington for legal data, the NSF Office of Science Information Services for human relations data, Project TALENT for psychological data, the National Opinion Research Center, for a continuous national survey of social indicators, the Ohio State University for labor economics data, the University of Michigan's Institute of Social Research for consumer satisfaction data, the University of Chicago for information on the behavior of securities markets, the University of Iowa for the Social

Science Data Archive, UCLA for biomedical data, and the community of museum scientists for data bases on the classification of reptiles.

As for shareable software, already widely used and shared are such statistical analysis packages as BIOMED, SPSS, and DATATEXT. Among the newer, innovative, and generally hard-to-transport packages are the "consistent" system of Project Cambridge at MIT; Project IMPRESS at Dartmouth and its daughter, POISSON, at North Carolina; the on-line commercial service of Data Resources, Inc., for econometric modeling, data retrieval, and analysis; the general-purpose file access and manipulative data language on ARPANET; and the Project Troll econometric modeling and analysis under development at the National Bureau of Economic Research's Cambridge Center.

Not to be omitted in a discussion of shareable resources is the basic computing resource, raw computing power. And then there are the most precious of scientific resources: ideas, enthusiasm, and intellectual stimuli, all inherently shareable. Some believe that one of the most important benefits of a network will be increased intellectual contact among dispersed scientists with common interests. The "invisible colleges" thus created or strengthened may have as much effect on the progress of science as the actual computations the network permits.

Classes of Network Applications

Within the foregoing examples are four types of network use that can conveniently be stated with the help of two definitions. By a *terminal* is meant a device like a typewriter terminal, a keyboard with a CRT display, a remote job-entry station, or even a minicomputer with limited memory. By a *processor* is meant a computer capable of handling (for example, subsetting, recoding, sorting) and performing statistical analyses on data sets of at least 10,000 to 100,000 words.

The four types of use, in increasing order of difficulty of the technical and organizational problems they raise, are the following:

1. A terminal connected to a processor, providing relatively standard services that could be obtained from any of a number of processors. Examples are access to BASIC for use by students in undergraduate courses, or to a computer with enough core memory to perform computations that cannot be done on the machine at the user's own college or university. This class of use is already common and is the backbone of several commercial services.

2. A terminal connected to a processor that runs a unique or unusual problem-oriented program that cannot in practice be transported to the

machine at the user's own institution. The remote use of IMPRESS for undergraduate instruction is one example; remote use of the biomedical systems on the IBM 360/91 at UCLA is another. The present technology of networks is quite adequate for this sort of application; what is needed is organization.

3. A processor, perhaps one the user has reached through a terminal, gaining access to a second processor that is custodian for a data base of significant size. Programs on the second processor must be invoked to extract the particular data that interests the user. This is technically much more challenging than the first two applications. If it is to be done automatically, without great labor by the user, it requires the first processor to command the operation of programs that run in the second, a technique that has not yet been widely developed.

4. A processor, again perhaps one the user has reached through a terminal, appealing to another processor either to perform computations beyond its power or to make use of special programs not in its library. The first processor must, in this case, command a much wider variety of programs than in the previous case. This is more difficult, and, indeed, approaches the bounds of what has been shown to be technically feasible.

Beyond these four classes of applications is the prospect of intelligent terminals with enough knowledge of the procedures and contents of the network programmed into them to help the user find his way through the system. This is a very important possibility, but the technical and organizational problems it poses are the most difficult of all.

Much of the benefit commonly expected from networks would really result from the organizational and administrative arrangement it could help bring about: documentation adequate for users at a distance; consultation with the curators of data banks; collection of programs fully tested; and catalogues of programs and data available from a variety of sources. If these arrangements were made anyway, in some situations, connecting to a network might *not* be worthwhile. If, for example, the speed of response is not crucial, large amounts of data might be transferred more cheaply by mail; some programs now available only at special locations could be made to run at other installations, and there would be advantages in doing so; and many of the intellectual contacts foreseen might still occur.

Network Problems and Liabilities

Potential problems should be addressed in the design stage of a network undertaking. Some are common to any computer operation; others are intensified by or peculiar to network operations. Some problems will affect all users of the network; others especially affect the social sicentist, data-base user, and modeler. The following major problem areas could have particular impact on the social scientist or data-base user.

Security and Accuracy
1. Programs and data must be protected adequately from accidental or deliberate change or destruction.
2. Some data bases must be shielded from certain users (or particular parts of some data bases must be made available only to certain categories of user). Restricting the availability of data protects its confidentiality while still permitting authorized persons to work with the data and take advantage of network facilities. But, by the same token, it does impose barriers to the use of the data, overhead in obtaining clearances, and the like.
3. Some data bases must be protected from exploitation for commercial purposes (that is, protection must be based on the intended use of the data rather than the identity of the user). Such requirements may well lead to restrictions on availability of data when viewed in the context of a national network.
4. The liability arising from breaches of security must be clarified.
5. Transmission accuracy (coupled with error detection and correction procedures) must provide a sufficiently high probability that the data actually processed and the answers actually received are the intended ones.

No new problems of security should be encountered in an existing network (like ARPANET) where each institution is responsible for its own security precautions. But a new network could have new problems. Networks may aggravate problems of security merely by involving more people and may degrade transmission accuracy by creating large increases in message traffic. One of the most important problems for biomedical and behavioral scientists will be protecting access to certain data bases because of their confidential nature.

Impact
1. The widespread availability of data bases may change the balance detrimentally between experimental replication and secondary analysis.
2. The increasing number of available data bases and application packages may cause a proliferation of data and programming errors unless effective certification and validation procedures are instituted.

3. Researchers, especially behavioral and biomedical scientists, tend to pick research topics or ask questions that lend themselves to treatment by available packages; increasing the number and variety of these packages may help remove present limitations, or it may only have the effect of channeling research in different ways.

4. The availability of specialized facilities via a network may assist those currently at a disadvantage with respect to services received, or it may instead provide a disproportionately large benefit to the already better-endowed researchers (that is, the rich may get richer).

5. Entrance requirements such as interface hardware and program contributions may work to restrict membership in the network and the network may gravitate toward serving particular classes of users (for examples, number crunchers and large institutions).

Documentation and Support

1. Documentation must be kept current either by the contributors or by another mechanism.
2. Documentation must be sufficiently clear, concise, and comprehensive to be of general use.
3. Support in the form of training classes, consulting, and question-answering services must be provided for network offerings by entrepreneurial-minded, service-oriented network staff.
4. The network may end up offering little more than raw computer power if it turns out that the imposition of documentation and support requirements discourage users from contributing their software to the network. Yet, unless programs and data bases are adequately supported, they are of little value.

These problems arise at the level of the local computer center, but they become far more important when user and producer are farther separated as in a network. Who will be responsible for a data base or for software after the original developer loses interest or changes institutional affiliation? Such questions need careful attention.

Member Institution Problems

1. An educational or research institution may not be able to accept an entrepreneurial role in providing computer services to a network. That is, such an institution may not be able to justify the acquisition of additional computer resources in order to supply larger quantities of a service that has become popular with other users outside the institution. The financial risk and the diversion of effort for the benefit of nonin-

stitutional users may be inconsonant with the institution's charter.

2. A "balance of payments" problem (inside users drawing more services from the outside than outside users draw from inside) may lead an institution to limit its members' access to the network.

3. An institution must have assurance that a satisfactory network service (upon which it might come to depend) will continue (a) to be available (b) at approximately the same cost (c) with the same response time and service quality. This is a loss-of-control issue to which institutions are very sensitive.

4. The provision of service to network users may lead to unacceptable rigidity for the host system. That is, an institution may be unable to shift to new and perhaps incompatible versions of an operating system (or to require new control cards and the like) when it desires because continuity of network service requires less-frequent changes and modifications.

5. There may be no incentive to make programs and data available to a network. The costs to document adequately, to consult with users, to guarantee continued availability and support may outweigh any possible gain to the custodian of the programs or data. Alternatively, the possible gain may have such a high risk associated with it as to be undesirable.

6. An institution must be able to estimate its costs and commitments before joining a network.

These difficulties may tend to discourage a potential user institution from joining a network. An institution that does not view a network as both safe and desirable has little reason to join.

The "balance of payments" problem may be particularly important to institutions with many biomedical or behavioral science users who tend to be more consumers than producers of service. For such institutions the network may threaten a net outflow of funds, which might cause the local computer center to run a deficit.

Network-Institution Relationship

1. The types of institutions and learned societies that are allowed to join the network, the conditions under which they may join, and the body responsible for determining their qualifications will have a significant impact on the character and function of the network.

2. Whether the network is active (controlling the offerings available to its members, assigning specialties to nodes to assure broad coverage, and prohibiting duplicate offerings) or passive (permitting anyone to offer

any service or facility) will similarly have a significant impact upon the character, utility, organization, and funding of the network.

3. Information must be obtained from and distributed to members on a timely basis regarding changes in offerings available (or conditions under which they are offered); in documentation, program operation, or data-base organization; in facility operational hours, rates, or procedures; and in control cards or system facilities affecting previously executable programs.

4. Procedures must be developed to enable a user to get clearance from his institution to use network facilities and to get an account number and password on another facility with a minimum of delay, effort, and red tape.

5. The responsibility for problems resulting from unannounced or improperly announced changes in programs and data bases must be established. Similarly, the responsibility for program bugs and bad data must be fixed.

6. Different pricing procedures for various network facilities (for example, costed versus noncosted) may pose problems for using institutions that are required to operate under particular accounting procedures.

7. The responsibility for providing general information about offerings, procedures, and the like must be appropriately divided between the network itself and the participating institutions.

8. It would be desirable to provide different aids or support levels for different users or different classes of users.

These problems, generally concerned with administration of the network and its interface with member institutions, should affect other classes of users as much as the behavioral and biomedical scientists.

Problems of the Network Itself

1. Some agency or body must be responsible for the operation and maintenance of the network. The long-run funding of this body will have a significant impact on the character of the network.

2. It might be desirable for the network to exist as a separate entity. That is, the network itself would be able to take over and support certain programs and data bases for general use when the originator could no longer support it. This is particularly desirable for the archiving of data bases and programs.

3. The network must be able to control its development (for example, to expand capacity as communications links approach saturation).

4. The network must establish reliability and performance standards.

In connection with this, there must be mechanisms to police and enforce the standards.

5. Whether the network merely makes available the offerings of participating institutions or actively markets service will greatly influence the type and extent of usage as well as the rate of its growth. The network's offering of training, documentation, or consulting services would have a similar influence.

6. A program or data-base certification procedure would be very desirable. However, cost considerations might dictate the use of graded offerings (for example, type 1 for those actively supported, type 2 for those made available at the user's risk, and so on).

7. It is important that the network have a means for self-evaluation so that momentum will not control the course of future development.

8. It is important that continuity (the continued offering of particular services) be guaranteed for users. This has implications for the organization of the network itself. It is also desirable to be able to persuade institutions to contribute and support materials on network terms rather than on their own terms.

9. The form of the network (for example, commercial enterprise, joint venture with member ownership, loose association) will have a significant impact upon the power and role of the network in the areas of performance, documentation, and standards.

The structure and form of the network administration is a crucial problem. Should the network be a separate institution with an independent existence, or should it be the sum of its member institutions? If the network evolves as a commercial system, it will have an entirely different set of goals and priorities than it will if its primary goal is to advance research and education. If the network is a joint venture, the legal responsibilities implied may discourage some institutions from membership. The structure of the network will be important in determining the way it grows: its membership, programs, systems, data bases, documentation, standards, and hardware.

Network Demonstrations and Tests

To evaluate the potential of a network for improving the quality or quantity of a researcher's work, a study of the activities, motivations, and existing "network knowledge" of a peer-selected panel of leading-edge scientists in some discipline should be conducted. The interviews and associated data collection transcriptions could be used to increase awareness of network potentials and problems, and to estimate the extent to which a researcher's activities would change if linked computers were available. As background for this study, there should be a systematic review of the discipline's use of existing networks for sharing computers, data bases, and software, and other systems of exchanging resources that lower total costs of research or facilitate the growth of the discipline. This review should limit itself to identifying the scientific accomplishments that were made possible through resource sharing, the most fruitful areas for continued sharing, and the relative advantages of computer sharing through networks and other kinds of sharing.

Information should be collected and distributed on a regular basis, preferably daily, about the use of a large data base by a large number of researchers. This effort should use all the preceding activities but should not be limited to them. The purposes would be

1. To provide on a daily basis information on newly discovered errors, missing records, faulty codes, or other anomalies (and current information on additions to common software since the last documentation update).
2. To set the stage for collection of information about new but related data files resulting from action on the main data base.
3. To collect and share information on costs of operation on the data base as fast as new techniques of cost reduction are discovered.
4. To test procedures for cataloging and documenting data files and data-base information in a multiuser system.
5. To test the willingness of researchers who stand to benefit from using a relatively simple network to share actively the technical byproducts of their research work via a computer network and to compare this to the current state of sharing via print media.
6. To test the problems of administering the relationships resulting from the combined sharing of software, statistical data, and document data.

It is also suggested that the following kinds of proposals for research grants or contracts be solicited from specialized research resources in the social or biological sciences:

1. Inward WATS. Fund inward calls for access to a remote facility by users wishing to test the feasibility of using the facility as though it were part of a network. The subsidy should include free usage to a specified level for several months to allow the case for continued use on a pay basis to be developed.

2. Outward WATS. Fund users to try various centers to evaluate the benefits and difficulties of remote use of the type that might later be

available through a network, including provisions for computer charges at the various facilities.

3. Higher-capacity line connection between a user group and a remote facility to allow a better grade of interaction.

4. Entry to ARPA. Fund the connection of two or more specialty centers with access to the ARPANET so their users can use the other facility and so they can serve a variety of ARPANET users.

5. Information center. Fund proposals by specialty facilities or groups to provide general user information on access to facilities, programs, and data sets of interest to a particular discipline. These groups could investigate the needs and costs for maintenance of data screening, analysis, or retrieval systems and undertake such maintenance in certain cases.

6. Network access broker. Fund applicants to develop the expertise to negotiate access to facilities that can best serve the needs of a selected class of users.

7. Training. Fund proposals for training consultants, applied programmers, managers of specialty centers, brokers, and user-group contacts for specialty facilities.

8. Documentation and other support. Fund proposals by established centers with data-handling-and-analysis package systems to improve and extend documentation, maintenance, and information dissemination on their generally used systems.

9. Reliability (backup). Fund proposals to explore how to provide backup for specialty centers serving users who suffer greatly from occasional delays.

Each of the proposals should indicate how it would help satisfy user needs for easy access, documentation, stability, information and consulting services, and file facilities.

Constituent Groups

A national computer network must, in the last analysis, have the strong support of some membership or constituency if it is to be viable. Based on the experience of single facilities and smaller nets, survival has frequently depended upon the interest (in terms of usage) and contribution (in terms of both financial sacrifice and intellectual development) of some identifiable group of individuals or institutions. This is doubly true in the case of any network significantly dependent on federal funds for its operations. Supporting agencies will back away from "data

morgues" or "software monuments" in view of the competitive pressures
on available money.

Because of its obvious structural complexity, a computer network
would attract many constituent groups rather than a single organization
of potential users. But not all these groups would prove to be strong
or effective. Limiting consideration to constituent groups that are contributors to, and not merely consumers of the services and materials
which the network can offer,* constituent groups fall into the following
classes:

1. Nonprofit educational institutions, including universities and colleges. They are likely to contribute major host facilities to the network
for use by scientists at other institutions, and they are likely to benefit
from the sharing of hardware and software resources residing at other
institutions.

2. Whole disciplines or branches within disciplines. In some disciplines,
computational methodologies have permeated all branches of the science,
while in others quantitative approaches have been adopted only in
certain sectors. The availability of the greater range of facilities and data
resources that a network can offer will appeal to these groups, and it is
they that will cause the network's conceptual and methodological growth.

3. Scientists engaged in basic research. This group already has a tradition of data bases and will have specific ideas for the kinds of facilities
and information required for the advancement of their fields.

4. Applied scientists such as those engaged in research on social problems (as contrasted to social science research). These scientists will be
drawn from all disciplines where extensive use is made of large data bases
and will be characterized by the view of networks as the facilitators of
rapid answers to classes of questions—questions addressed to existing
knowledge rather than those that require the generation of new knowledge. This group, along with the basic scientists, will be important
contributors of new data files and programs.

5. Scientific groups, teams, or organizations that define their research
activities by area rather than discipline. These are the interdisciplinary
groups set up to work on broad problem areas, or "invisible colleges"

---

* For example, faculty members at small colleges who, under current NSF programs,
can attach themselves to ongoing research projects at major academic institutions
and, for limited periods of time, draw upon the resources available to such projects—
including computer usage—would not, despite their large number and potential in
other regards, be considered a constituent network group because they are unlikely to
contribute resources of any consequence.

of scientists bound together in a common cause to produce some scientific development in the near term. They are likely to be strongly interested in and to benefit from the sharing of new knowledge and advances in technique made by colleagues, as well as to appreciate the practical advantages of utilizing data and programs developed elsewhere. They are also likely to contribute to the further development of new data files and programs.

6. Existing facilities and protonetworks. Good examples already exist for groups living in prototype network environments where the benefits of further extension are likely to make the most sense. Examples include museum-based data banks containing information about important systematic collections from archaeology to zoology and specialized social science collections of survey data and aggregate statistics.

7. Funding agencies. Groups of scientists that recognize the interest of outside supporting agencies in introducing computing economies or promoting scientific advance may be expected to form as a natural result of competition for scarce funds and interest in certain specific kinds of scientific work.

These groups will, of course, be made up of overlapping memberships. Individual scientists do not and will not fall neatly into mutually exclusive clusters.

References

1. John LeGates, "The Lessons of EIN," *EDUCOM Bulletin of the Interuniversity Communications Council*, Vol. 7, No. 2, Summer 1972, pp. 18–20.

2. Hugh F. Cline, "Social Science Computing, 1967–1972," *Proceedings of the 1972 AFIPS Spring Joint Computer Conference*, AFIPS Press, Montvale, N.J., 1972, pp. 865–874.

3. George Sadowsky, "Future Developments in Social Science Computing," *Proceedings of the 1972 AFIPS Spring Joint Computer Conference*, AFIPS Press, Montvale, N.J., 1972, pp. 875–884.

HERBERT M. TEAGER
Boston University Medical Center

15

The Exotic Medical User and the Ongoing Computer Revolution

The purpose of this paper is to view the unique needs of the medical research and clinical care community within the framework of a proposed computer network. As of a decade ago, it might have been sufficient to spell out needs, accepting future system characteristics on faith and ignoring costs. Today such an approach would be not only naïve but probably detrimental to the early realization of our particular societal goals.

Lessons from Past Experience

The odds against a successful network are far too high to go unexamined. While it may come as a distinct shock to the regular reader of computer publications, computer systems fail with dismaying frequency to meet their goals at all levels. Far from there being a lively professional interest in autopsies on such efforts in order to look for hidden pitfalls and avoid repeating past mistakes, they are seemingly papered over by the next wave of exciting new approaches, which are usually heavily larded with the newest generation of hardware. So long as computerizing efforts were research oriented, the only potential risks were inexhaustible government dollars and professional reputations for prophecy (and by tacit consent even these suffered little). Over the past decade, however, computer systems have increasingly intruded on all aspects of our society, including decision making on a national level, and the tragic consequences of computerized blunders are all too painful to the general public consciousness to be ignored by computer professionals. Further, our increasingly alienated and depersonalized society is ever more leery of science and technology and is no longer in the mood to buy without intense scrutiny just any scientific project. National resources *are* limited, and thus both cost and effectiveness must be very carefully weighed by systems designers rather than vaguely heralded in the initial press releases. Future failures, even if suppressed, can only act to harden attitudes further. Thus, if networking is to be the next big computer push, it must bear responsibility not only for its own viability but for the continuance of applications it is using for self-justification.

Since computer system failures go generally unacknowledged and unremarked, it is hardly surprising that no systematic efforts have been under-taken to find out where we are vulnerable to future fiascoes. Let me submit that there is nothing mystical in uncovering our weak spots and taking preventive action. The solution lies in changing those aspects of the behavior patterns of academia that are self-reinforcing and inherently unstable and allowing for healthy corrective negative feedback at all stages, including initial devil's advocacy.

Large system projects are the hallmark of our society, and computer efforts in particular suffer from the mystique. It seems to have been the general belief ever since the World War II Manhattan Project that success in technological endeavors requires large numbers of academic people arrayed in a critical mass, centralization of both evaluation and control, exploitation of a single technology, and the allocation of huge resources. The classic case of system failure, namely the World War II naval torpedo, bore all these approved stigmata, as did its more com-puterized descendants like the Air Force command control systems and the defense management systems. It also shared additional properties such as self-evaluation, a captive, beholden user group, and hubris, all of which were duly noted by the Navy before they literally destroyed all traces of the group that had unheedingly saddled it with an ostensibly advanced but tragically erratic and malfunctioning torpedo for two critical war years. In terms of exposed failure and deserved denouement, the torpedo is unmatched; today's more sophisticated organizations can usually be counted on to redefine both costs and effectiveness as they proceed, and even in the case of gross, catastrophic misfortune, appro-priately orchestrated public relations can muffle the crash until time can make us forget it. The unmarked casualities are the initial objectives that were to be met with such éclat. On subsequent attempts these are dropped as proved impossibilities or, equally disconcerting, are dismissed as irrelevant because the project has turned to educational activities.

Within the framework of the large, sole-source systems project as it is now constituted lies a fundamental inability to entertain initial alterna-tives, to evaluate progress without bias, or honestly to admit and correct errors or prevent their institutionalization. Before a project starts, there are strong pressures to justify the effort by its assumed obvious benefits, and once the project is under way too many vested interests are at stake to allow the intellectual honesty of appraisal and admission. Even the process of setting up an academic systems project is fraught with pressures for short-sightedness and conformity under the guise of prag-matism. A summer study or equivalent forum is generally convened as a means of generating consensus for a preexisting plan rather than as a means for evolving and critically examining one. Much of the working time of such forums is spent in seeking out subtle clues on who is in favor in the prevailing Washington climate and thus how best to package the project. The fact that such systems will then be judged by more senior, experienced members of the same club who are temporarily stationed in Washington helps to keep the doors impervious to disturbing notions, ideas, and people. The wisdom, let alone propriety, of this form of academic incest is open to question if one has a less rosy view of the success rate of past computer efforts. It is further not at all surprising that senior advisers stray very little from the confines of their increasingly tight little disciplines.

In all fairness, computer projects are seldom espoused for the overt aggrandizement of the computer fraternity. Rather, they are always touted as providing high-order intellectual services to some deserving group. A fascinating question is how captive users are kept informed as to what is possible, yet free and unbiased in their appraisal of the quality of services provided, considering their dependence on the providers. The answer would appear to be the obvious one: users are neither free, informed, nor unbiased.

So far we have merely suggested that large-scale computer systems are subject to pitfalls that are similar to other high-technology systems. There are, however, additional traps related to the thin analogy between computer information processing and intellectual activities that make success even more elusive.

Some of our more poetic colleagues have pointed to the far-reaching intellectual impact of the invention of the computer, ranking it with the ax and the wheel. Certainly if the acronyms chosen for computer systems —"Prophet," "Oracle," "Artificial Intelligence," "Machine Aided Cognition"—represented performance, the millennium would long since have arrived. Unfortunately, computer technology for the past twenty years of my experience has been besieged by solutions in search of problems, imminent breakthroughs awaiting the next order-of-magnitude increase in memory, speed, or budget, and a plethora of ham-handed man-machine interfaces whose designers seemingly wished that people were more machinelike. Even hardware, particularly at the end of telephone lines, has shown a strong desire to frustrate and mangle. We seem to know not a whit more about man-machine interaction than we did a decade ago; every *major* intellectual task that, according to the literature then, was within our grasp now seems to have been tacitly proved impossible or to require unlimited additional basic research. Supporting data on critical parameters, problem definitions, or cost effectiveness issues are almost impossible to find.

The intervening years have, by and large, been good to computer science academia. Masses of papers have been written, courses added to the curriculum, the pipelines filled with more Ph.D.'s, and computers have churned from their remote accesses. Perhaps the greatest outcome of the computer revolution has been to mechanize bureaucracy and substitute printouts for thought to a degree that even C. Northcote Parkinson has yet to realize. One would have to be blind not to note even the more benign of computer horrors, like automated squad car scheduling, classroom scheduling, crime statistics reports, computer-aided design of power plants and unwanted highways, and various management information systems. Lest we forget, Norbert Wiener had a few choice words to say about the potentially evil uses of our favorite toys, and it might be well to consider the far-reaching effects on a nation's morality of that simple computer metric, the "body count." If we as a field have not yet known sin, it is perhaps because we would rather not look. It is indeed "our department where they come down."

It would be interesting some day to explore the ease with which so many of our technical and political leaders have utilized computer science to advance to ever higher positions while leaving a path of havoc and disaster in their wake. Perhaps it has something to do with time constants for failure to materialize, coupled with our new systems engineering predictive techniques for proclaiming proof of success. Then again, the phenomenon may not be new after all—the snake oil salesmen of the last century also had instant cures and optimized schedules.

If the field has been intellectually dishonest in the past, it is not because a few misguided cranks did not try to insist that scientific, or at least engineering, standards had to be applied to the art form of computing. When some of our more illustrious research projects announced results in their proposals and national meetings, more of the fraternity could have protested the constructing of large systems with only vague justification and insisted that both cost and effectiveness had to be recognized as the important issues.

All the foregoing brings us to the question of our latest white hope, networking. One must first of all ask in all candor just what might be available at a cross-country node that is worth having, and second, what are the odds that networking can provide it in a usable, reliable form and at an acceptable long-term cost. These are not trivial questions. In looking at the ARPANET prototype, my first instinct is to ask what it has produced in the way of tangible, useful results. As has already been suggested, an unbiased observer in this field is somewhat of a rarity, so that potentially self-serving conclusions deserve a bit of scrutiny. The papers that have appeared on this subject have all assumed that there was in fact something worth sharing at each network node and have spent most of their time plugging a particular brand of data-packet shipment. In terms of the technique, it seems worthwhile to note that the data rates involved, while seemingly high, preclude the shipment of video image data and further require a reasonably large terminal invest-ment to tap into what can be shipped.

With these provisos in mind, let us now consider real-time medical applications that possibly, if the services provided were appropriate, re-quire a network hookup. Thus we will restrict our view to examples that might be allied to national information bases, specialized, one-of-a-kind processors, or unique, nontransportable programs.

The Biomedical Community

My official title is professor of medicine and chief of biomedical engi-neering of Boston University Medical Center. As such, I am charged with bringing modern technology to basic biological research, clinical research, and the delivery of health care while existing in the bureau-cracy of a large university and medical center. By that definition I find it difficult to do my job. I have been at it full time for six years and off and on for the prior decade. I have unfortunately lost most of my com-forting illusions. I know too much to survive getting sick—or at least I have to forswear the beneficial belief in medical omniscience. Medical debugging, or patient care, has all the heuristic certitude of Russian roulette, and practitioners are no less fallible than a beginning pro-grammer.

In my laboratory we have operational amps, HP pocket calculators, several types of minicomputers, a medium-size PDP system, a resuscitated weapons system (A-New Mod 0 to be exact), an IBM 360/40, and a teletype-dataphone link to the main campus computers. Looming on the horizon is a high-speed data link to the main campus system's IBM 370/175. Each of the foregoing is connected to its own unique or pooled set of peripherals, from image and sound processors to tapes, cards, cas-settes, printers, and displays. One of our most impressive systems swings around four 500-pound nuclear devices (scanning heads) for a radioiso-tope scanner, while another runs a gas sampling device and gas chromatograph of our design in an intensive care, cardiopulmonary patient monitoring system. If you inquire about almost any kind of computation, information retrieval, instrument control, signal analysis, real-time interaction, or on-line experiment, the odds are high that we or one of our colleagues are doing it. How well we are doing it is another matter.

For intensive computations, we are embarking on simulating nonlinear, transient, hydrodynamic phenomena occurring in living systems and organs. For extensive work, our epidemiologists, community health, and psychiatric researchers are collecting and processing ever-larger data bases to find disease correlations and deliver community health care. But let me now be a bit more selective and pick some user examples where the issue might be more directly allied to national information bases, specialized one-of-a-kind parallel or pipeline processors, unique programs, or unique distant programs to which local access is required.

Patient Records

The largest potentially useful medical application is one where major chunks of design data are missing, clinical care. The prototype patient data base is the current patient record, which unfortunately is the major source of information on which treatment is planned and followed. The patient record is inadequate on the grounds of omission, commission, and accessibility. Instead of photographs and x rays (which, while available at local depots, do not fit the manual format, let alone the digital one), it contains short notes and summaries from specialists. Written and image material lies inaccessibly scattered at every clinic or doctor's office to which the patient has had contact during his possibly peripatetic career.

Since each of us is unique, self-comparative data is vital. The radiologist is grateful for whatever bilateral symmetry he can find. It is lucky that we have comparatively few "single" organs and that simultaneous bilateral involvement is rare. Time-series data and time-spaced images of the same organ are likewise important. But too often these data reside with different physicians in different places.

Intrinsically, the patient record of the average sick patient can run the size of the old Brooklyn telephone book. Attending the patient may be some fifty residents, surgeons, internists, and varied specialists, nurses, therapists, and aides. The patient record does not help their interpersonal communication, nor does it facilitate the presentation of laboratory data, vital sign charts, and electrocardiograms. Each painfully extracted sample is transmuted into an easily filed slip of numbers in the yellow pages in appropriate drugs and dosages. Much out-of-line physiological data and values and incorrect assumptions go unseen and thus uncorrected.

The system obviously can be improved. Computers are part of the cure, but we also need to be able to combine image data with the patient record. Microreproduction and high-resolution video are good enough. Perhaps eventually we should constrain our doctors to think more quantitatively with interactive digital records, coded comments, and machine diagnoses, but a system that could circumvent data lags and make all currently available information accessible to a bedside would greatly enhance diagnostic acumen.

The choice of input-output equipment, however, represents a touchy problem in man-machine matching. For example, keyboards and doctors do not mix, so input devices had better accept handwriting. Finally, any computer-based improvement had better cost less than $3,000 per bed, or else forget it. Hospitals are not the Defense Department.

Consider now a civilian who is found unconscious in a gutter. If lucky, he may have his wallet intact, be white, and appear well dressed. Even assuming that he is swiftly transported to a center of mercy and his diagnosis is obvious, the probability of a specific treatment being bad (possibly fatal) for this particular victim is too high for comfort. The facts that easily might be nestled safely in a folder a few hundred yards from the patient are not going to help him!

Most engineers know that a 110-volt shock will only jolt them, but for 5 percent of the population it can be a killer. The same applies to drug sensitivities and differential diagnoses that go the wrong way because of the unavailable past history. But before we all jump after an accessible medical history system, let us not overlook the damage that can be done with the leakage of medical information. Basic decency prevents much of what is known about most of us from becoming public; but we seldom hear the telling person-to-person whisper.

Patient Care

The foremost design constraint of any intensive care system is that it must be trustworthy, which means it has to be reliable, self-checking, and failproof. It goes without saying that it should be of benefit to the patient. Other factors that must be considered are the importance of the basic parameter, measurement accuracy, the risk to the patient, noise, distraction, hazard, incitement of fear, accessibility of derived parameters, ease of emergency changes, numbers of personnel involved, and ultimately, cost. The dollar cost merely reflects the resources, and even if equipment were no object, the amount of short-supplied, highly trained specialists would be limiting. *The Andromeda Strain* in the movie version portrayed a lovely ideal: uncluttered, pleasant rooms; no extra holes or appurtenances in the patient with wires, tubes, and bulky transducers; no medical collection system for his vital juices; an unharried, loving staff that can go to a retrieval system and refresh their memory; and a complete up-to-the-minute status of all patients' systems and treatments—complete with go–no-go limits. Would it were true, or even in sight, or possible!

In trying to acquire a suitable gas sampling and composition measuring system from the industrial suppliers for assessing a patient's pulmonary status, we could not find a single gas chromatograph (our choice for the analytical instrument) that came close to a decent set of design specifications (their sensing elements burned out on an enriched oxygen supply, the sine qua non of respiration therapy). The unit coming closest—just the analytical instruments, not the necessary sampling system—went for $20,000 plus full development costs for modifications to put it into an automated on-line system. We ended up building our own computer-controlled chromatograph and sampling system. It cost us about $10,000 and a year of grief, but we are in reasonable shape on at least one set of parameters. Now we want to automate a respirator, and we are shuddering at the inelegance of mounting servodrives and shaft encoders to the ten control knobs on the machine; but perhaps someone else will do it first, and save us the shame.

Then there is the blood, gas, and electrolyte machine to which we are going to interface. Such standard and easily interfaced or automated devices as thermal and dye dilution meters (blood flow), pneumotachs (respiratory gas flow), EKG electrodes, blood pressure catheters and transducers, or even radioactive scanners do not yet present any bad problems in cost or reliability. Calibration, preferably fully automated, and malfunction detection can be difficult. Moreover, we cross-check each measurement with a completely independent technique. It is embarrassing to have a machine decide that a patient is hors de combat just because his EKG lead or (far more serious) arterial line broke.

There are at least four real-time (or nearly so) areas with a strong and growing computer component: coronary diagnosis and monitoring, radiation therapy (planning and administration), clinical laboratory automation, and radioisotope scanning. The medical consumer is again following tradition in settling for the first system that can achieve a desired function at a subastronomic budget, as devised by computer and instrument manufacturers with a vested interest in selling the most expensive and elaborate systems they can put together and get field-tested. Systems that should be possible to build for $5,000 to $10,000 are sprouting like mushrooms in the $100,000 to $200,000 range. Since most of the software is proprietary, there is no immediate glory in reinventing the wheel or deciphering the Rosetta stone, and the essential preliminaries to splitting off data collection, data storage, information processing, and real-time control functions into hierarchies go undone. I suspect that when the day comes to tie these systems together locally our options will be either a $20,000 tape unit or a $10,000 modem per unit.

Medical Research
Finally we come to medical, surgical and physiological research, which is where the ultimate differential between mere treatment and prevention or cure lies, as well as Nobel prizes, fame, fortune, and the covers of *Time* and *Life Magazines!* The key issue in medical research today is *biochemistry,* not electronics. But increasingly, our biochemical instrumentation (such as chromatographs, microscopes, and spectrum analyzers of all kinds) is becoming automated and equipped with inexpensive minicomputers plus expensive peripherals. We have done enough building and looking ourselves to be reasonably sure that the same happy hunting ground for the OEM (Original Equipment Manufacturers) applies here as in the case of clinically oriented devices.

Two information areas universal to all research efforts at my medical center are data analysis and literature search. Data analysis, except for epidemiologists who have other problems in firmly establishing ex post facto cause and effect, tends to be a matter of stretching small sample statistics to the limit. Besides, setting up physiological experiments to produce a chart can be so horrendously difficult and intrinsically messy that one tends not to worry too much about experimental design or the precision of the results.

Finally, let us consider the problem of the casual or serious browser wading through the morass of published literature. Occasionally, to answer a question for which a survey has not appeared in *Science* or *Scientific American,* one must brave the literature and, as a last resort, the modern data-bank retrieval system. Of course, the best retrieval system known is the informed colleague.

I have had occasion to use one of the most modern of the computer retrieval systems recently and came away totally dismayed at how such a potentially marvelous idea could have been so badly handled. First, I had to make friends with the young lady operating the system, even to employ her as a sound-to-teletype keyboard transducer. The usual dialing, password, and other sign-on incantations were completed, since she was good humored as well as pretty and smart. But the system will soon fix that! Either she will become ill-tempered or the library will find a harpy to stand the gaff. Second, my "buzz" words elicited a "ho-hum, never heard of you" response from the system until I retreated to higher, safer, and more all-inclusive descriptors. Third, it turned out that the data were classified or confined like a prisoner under the Geneva War Convention by name, journal, title, descriptor list, next name, and so on. What it did not answer was what I most wanted: a summary of the article and its citations. Also I was disappointed by the enormous omissions of domestic, let alone foreign, work. Finally, I had to resort to an off-line listing, followed by Xerox copies of half of the papers, in order to locate two of interest out of the original one hundred—requested ten at a time, ten times. Where, oh where have microfilmed indexes and journals gone?

In the area of mathematical modeling, although precious little of vital concern is known about the multiple, nonlinear nervous, hormonal, and humorous control systems of the simplest organs, we shall continue to press on with our linear system models.

Few exotic areas of computer technology fail to find sympathetic vibrations in the world of the medical center. General interaction, automated teaching, language translation, automated diagnosis, scheduling, resource allocation, pattern recognition, and sensory perception are all fair game, and they share all the flaws of the other areas. In the health field, the temptation is to yell "Your money or your life" and to forget that our biggest stumbling block is our own sheer ignorance.

Networking Requirements
Finally, let me attempt to be more explicit about the requirements of my kind of potential network user.
- We need quick, secure, useful access to national data banks. Unfortunately, cost is not the only problem, and digital data are not the only requirement.
- We would like cheap large-scale computation with short access and and running times, but we are less likely to succeed than the Weather Service on such a vast enterprise—and it knows what equations to use.
- Our real-time control problems tend to be of the sort where a busy signal, lost data, or bad data would be worse than a blunder, and money *is* an object.

- There is not any great outcry for a one-of-a-kind program or data that are available only at the West Coast, at least not at current communication prices.
- By the time we know our problem well enough to do a first-rate breakdown into a hierarchy of software and hardware of differing cost effectiveness, everyone has either lost interest or is willing to pay dearly for the prototype manufacturer-produced version.
- In an era where we have to fight for pennies, let us invest a few dollars in raising the state of the art before hopping on the latest bandwagon, or at least let us justify networking on someone else's requirements.

Conclusion
When all is said and done, past organizational and technological fiascoes are not likely to influence the future. Ready or not, we are in for networking and our current misfortunes will end up frozen in concrete.

The largest pitfalls lie in a misappreciation of human nature. The tragedy is that it is really not so difficult to provide feedback and safeguards for the user in whose name systems are built. The basic questions are how to select broadly disciplined, creative, and knowledgeable system designers who are, and stay, honest and responsive to user needs; how to support a user population that has every incentive to demand the most effective tools that can be devised at a reasonable price without fear; how to leave the door open for necessary future development; and how to allocate scarce resources to prevent stifling alternatives.

There are several keys to the situation:

1. Remove the temptation for instant glory by setting up long-term funding with no incentive pay for the government administrative agency (or contracting university) to make the project as large as possible, to increase the numbers of professional people involved, or to issue declarations of instant success.

2. Allow for intellectual competition. There has been enough waste in the use of sole sources. Set up at least two alternative approaches for each function to be accomplished at the network nodes. For example, make an honest try at program and data sharing before drawing the conclusion that facilities are truly so unique that only networking will work. Further, subsidize efforts (similar to the current mechanized microfiche Sears Roebuck parts catalog) in the local information retrieval business.

3. Allow for further development of allied technologies (such as high-resolution video, microimage storage, and computer access of image stores) before accepting today's data transmission techniques and con-

straints as immutable. If networking uses the ARPANET technology, the door is permanently sealed against image transmission.

4. Provide every user with access to an independent ombudsman panel of citizenry without an ax to grind charged with realistic evaluation of the system's development and alternative modes.

5. Since computers have generally made little positive impact on C. P. Snow's other culture, entertain at least one imaginative proposal from a university to integrate fully computer and other ancillary technologies into the intellectual life of the university, with appropriate safeguards against the nest-building and parochial tendencies of computer scientists.

6. Before being carried away by the potential benefits of unlimited access and sharing, take realistic note of how seldom one researcher is really willing to give up his treasured source data to a competitor and how often in today's straitened circumstances a computer center is willing (at less than usurious rates) to let loose their treasured programs.

7. Establish a mechanism for user identification and security that goes beyond today's password. Handwriting input, for example, is only generally feasible with a known user.

It is a crime to let a major (perhaps last) chance to make positive use of modern technology go to waste and poison the well once again. A country that can abandon interplanetary exploration is clearly not in the mood for a repeat performance of the blunders of the sixties. Finally, it is well to remember that sophisticated or not, man is a sapient being. Devices such as the HP pocket calculator neither existed nor were predicted ten years ago in the heyday of time sharing, yet today they are proliferating like wild flowers after a rainstorm. If a device is needed, and appears in a usable form, it will be utilized until something better appears. We should be meeting needs rather than finding usages for techniques. In that sense, the problem is engineering in the highest sense rather than management, mathematics, or computer science.

16

Interactive On-Line Responsive Systems
Report of Workshop 3

JAMES L. MC KENNEY
Harvard Graduate School
of Business Administration
Faculty Discussion Leader

Contributors:

BRUCE K. ALCORN
Western Institute for
Science and Technology

ROBERT L. ASHENHURST
University of Chicago

MAX BEERÉ
Tymshare, Inc.

C. VICTOR BUNDERSON
Brigham Young University

DONALD L. HENDERSON
Mankato State College

BEVERLY HUNTER
Human Resources
Research Organization

THOMAS F. KIMES
Austin College

J. C. R. LICKLIDER
Massachusetts Institute
of Technology

TERRENCE W. PRATT
University of Texas

LYLE B. SMITH
University of Colorado

HERBERT M. TEAGER
Boston University
Medical Center

KARL L. ZINN
University of Michigan

The initial focus of discussion among the participants in the workshop was on the reasons for its existence. Why have a group to formulate ideas for a paradigm of on-line real-time computing? It was concluded that the topics the group would consider were, in part, problems of other working groups but included a unique set of issues for both computer system design and user characteristics which warranted detailed exploration. The other workshops would cover such aspects as on-line library systems and modeling in the data-analysis systems, but they would do so from the point of view of the researcher versus the general user.

The characteristics of the users this workshop would consider—access to computer systems that have a high degree of reliability and a rich variety of software—warranted exploration of the broad class of users of on-line responsive systems.

A defining characteristic of on-line responsive systems is that the user interacts in real time, that is, the real-time response must be adequate for the particular task. In process control of medical experiments this could be microseconds; in an educational setting it might merely mean access to the computing resource before a specific date for an assignment. In demonstrating models on a terminal to a class of participants, a thirty-second response might satisfy the need. Emphasis was put on the breadth of the needs of the users to be considered, who include registrars, behavioral researchers, medical practitioners, and all varieties of students. To explore this range of needs, a reasonable agenda seemed to be to develop a taxonomy of users and then to divide into small groups to analyze their specific needs.

Use of Computers in Education

The participants of the workshop had a strong interest in the educational use of computer systems in colleges and universities. Further, several members were interested in the controlled growth of computer usage in their own institutions or in marketing the use of computers in other institutions. Thus, a recurring theme of the discussions was how best to manage the usage of computer systems as a tool for learning. In particular, a common theme was that such computer systems would have to be responsive to the particular student needs of each institution.

This led to a discussion of computer-supported education. A strong difference of opinion existed as to the necessary form of developmental support needed to exploit existing computer systems for educational purposes. One group argued that without a significant change in resource

allocation for development a more comprehensive, well-designed, computer-supported educational program was not attainable. Such a program
would require a costly developmental effort in order to develop high-
quality, economical education. The other point of view was that such a
complex educational system must evolve from the use of computing as an
adjunct to the classroom over an extended period of time. The process
of evolution would need continuous expenditure on computing but not a
massive investment. Such evolutionary development includes readily
available computer support and a range of prosthetic devices to induce
faculty members with a strong bent for teaching to become involved. It
assumes that if they bring their teaching skills to the computer and
learn to use it in pursuing their educational goals, a comprehensive system will result. This process would emphasize the computer's availability
as an adjunct to courses and would rely on its firing the imagination of
instructors on ways to use it in their courses. There was some feeling that
such a change was now taking place and that in some institutions
computers were being used in comprehensive main-line instruction.

This discussion of educational uses identified two ends of a spectrum of
computer-aided instruction: one, adjunct; the other, targeted. In the
adjunct type, the computer system is a useful but not necessary aid to an
ongoing course. Often it serves the function of a laboratory for particular objectives it alone can provide, but it is rarely a vital element to the
educational process. The targeted type would use the computer system
as an essential part of the program. It would be developed by instructional psychologists and dedicated educators and would use a broad range
of computer-related technology.

Taxonomy of Users
A classification of users is proposed that encompasses present and potential users of on-line responsive systems. The classes range from one that
includes the bulk of educational on-line computing today to one that is
untapped but that has an extremely high potential. These classifications
are (1) problem solving, (2) design activities, (3) simulation and modeling, (4) data acquisition and process control, and (5) instructional systems.
This classification was made on the basis of grouping common, on-line
user needs. This is an ambiguous issue, as can be seen in Table 16.1,
which shows how the various working groups identified their user needs
and important system characteristics.

The particular requirements identified for on-line problem-solving users

Table 16.1 User Needs and Important Characteristics of Various Working Groups

Problem Solving	Design and Editing	Simulation and Modeling	Data Access/ Process Control	Instructional Systems
User Needs or Characteristics				
	Engineer, Author, Musician			
1. Naïve—Canned programs	1. Knowledge of application	1. Advanced under-graduate or graduate —knows computer —uses as tool (lab)	1. On-line data collection	1. Large groups of students —geographically spread —common needs
2. Minimal programming —one high-level language	2. Sophisticated computer user	2. Faculty in research	2. Automatic control of experiments	2. Educator designing systems
3. Sophisticated programming—many high level	3. Relies upon computer as tool	3. Faculty in teaching	3. Human mediated control	3. Developers of instruction
4. System programmer			4. Ancillary analysis	4. Learners outside instructions
			5. Follow-up activity	5. Administrators of school systems

Problem Solving	Design and Editing	Simulation and Modeling	Data Access/Process Control	Instructional Systems
System Requirements to Meet Needs				
1. High reliability	1. Integrate software from different sources	1. Special-purpose modeling language	1. Proximity control	1. Special language for course developers
2. Good diagnostics	2. Access to multiple designs or copies of work	2. On-line system documentation	2. Availability of backup system	2. Audio, video color, and hard-copy synchronized control
3. Fast compiler	3. Capable console for author/designer	3. Reliable data management system	3. Allowance for human response time	3. Administrative system for student aid
4. Hard copy	4. Backup computing capacity	4. Access to analytical programs	4. Close tolerance control	4. Documentation system for instruction systems
5. On-line instruction		5. Flexible in/out systems	5. Auxiliary storage	5. Demonstration mode for showing off system
6. Help system		6. System stable over long period of time	6. Hierarchy of computers	
7. Multiple languages				

seemed useful for all user groups. The problem-solving class of user includes the bulk of those who now use on-line computing as an adjunct in higher education. Their objectives include (1) learning about computers or introducing programming languages, (2) accessing canned algorithms to solve real problems with large computations, (3) using the interaction of a computer and data to develop heuristically a solution to a complex analytical problem, and (4) learning a language such as French with an on-line programmed instruction system.

In general, an on-line system is used to provide prompt, comprehensive feedback to a user in a relatively complex problem situation. The problems are broadly defined and typically are concerned with some learning experience. The user comes to the console with a specific set of goals and relies on the power of the computer to give him information and response to a set of queries. The objective of the user is to leave the console with a better understanding of the problem.

The second class of user is those who would use an on-line system as a design aid. Its members have a specific problem in mind and turn to the computer for its ability to reproduce complex text and figures, large storage capacity, codified methods for reducing drudgery, and quick response, with a complete articulation of a set of summary inputs. These users could be editing text for a book, designing a bridge, creating a play, or writing music. The user has a specific product he wishes to create and relies on the system to generate an expanded version of his input, relying on previously articulated protocols, formulas, vocabulary, or grammatical rules. Further, the user may wish to put together a range of different software packages for a particular job. Where appropriate, he may choose to invent new protocols, shortcuts, and rules so that the next time he comes to invent or design something he can have access to their aid.

The third class of user is those employing on-line systems to develop and test models or abstractions of real-life situations. The testing process typically involves a simulation of some real-world system like a river, an inventory system, or a transportation system. The difference between the second group and the third is that the latter attempts to abstract a set of parameters and relations for testing some aspect of the environment. He depends on the computer to have a set of subroutines or prepackaged procedures to facilitate model building. Further, there should be input-output devices that permit him to enter quickly a few easily interpreted clues as to what he wants the model to do; upon testing, the output should be quickly analyzable for further testing.

The fourth class is those whose concerns are for acquiring data in real-time or controlling processes. The main users may well be researchers interested in acquiring information from such things as an acceleration or a medical operation or a functioning biological system. The group felt that quite often these systems might best be served by a computing hierarchy—a small on-line computing system acquiring data or monitoring the process and transforming data or appropriate signals for more extensive on-line processing by another system. The user's time dependency differs with the particular task. Thus, on-line surgery can use a slower response rate than on-line monitoring of a bubble chamber, although the rate of data transfer in both might be extremely high.

The final class of user includes those designing, implementing, and using an instructional system. An instructional system involves the student in the initiation of his or her own learning experience. It has a wide capacity to respond and anticipate student-initiated activities such as information search, performing statistical analyses, assessing his progress in learning; requesting and participating in tutorial dialogues, exercising a model; practicing basic skills. The purpose of the system is to deliver to the student a well-engineered, adaptive learning environment that can tailor itself to the student's particular needs and allow growth to proceed at the most effective rate.

The Identification of User Needs

The workshop was split into five groups, corresponding to the five classes of users. Each group was asked to consider the user's objectives, how a network could help attain these objectives more easily, what the critical problems are in developing networks or resources among the users, the key policy issues one must consider in developing a truly service-oriented system, and experiments the National Science Program could implement to achieve a better understanding of the issues or problems.

The workshop groups approached the challenge from a range of perspectives. For the intent of this report it would seem important to describe the user as perceived by each of these particular work groups and to discuss the capabilities he needs in the system to accomplish his objective. The system requirements are speculative and are defined in the spirit of what such a user would really like to have. Table 16.1 identifies the user needs or characteristics and the system requirements as defined by two or three experienced computer users. At times they would seem to be a Santa Claus list, and that in part was their intent.

On-Line Problem-Solving Needs

In the problem-solving classification we see a continuum of users that can be divided into four levels. The bulk of these applications represents the present form of adjunct computer-supported learning.

The first level is composed of naïve users relying solely on canned programs to solve their problems. This might be a student of French, or a student testing a broad range of algorithms, or an accounting student trying to understand the cash flow of a firm.

The second level of user consists of those with a minimal knowledge of a high-level language. He is a good conversational FORTRAN programmer intent on modifying a canned algorithm to fit a particular problem definition or on taking a data array as generated from a linear programming algorithm and making it amenable to a statistical regression program. Perhaps he might even develop a small FORTRAN program to help him solve his income tax or, on a specific assignment, link together a series of subroutines and print out a graph of the data within the FORTRAN, BASIC, or ALGOL program.

The third level would be users with a sophisticated knowledge of a high-level language. In a laboratory, one would be expected to take a complex set of differential equations and produce a fairly concise program to generate a solution. Perhaps he is conversant in APL and intends to generate a whole set of interlocking subroutines for use in a mathematical programming course. He may be a student in a graduate program who intends to take the Standard & Poor's statistics on industries and perform statistical analysis within the FORTRAN language system.

The fourth level of user contains sophisticated programmers, typically either developing canned programs or teaching aids for the other users, or faced with unusual problems such as writing a management system for the census tapes or developing a program for the analysis of text. This user requires complete access to the system and, in fact, may be helping design it.

Design User Needs

The users concerned with computer-supported design and editing must be qualified in the use of systems. A user must be an experienced practitioner of the trade, for example, writing a book or designing a bridge. He must also be reasonably sophisticated about computers and must understand the basic aspects of formal protocol, programmed instruction, and the value of labeling, indexing, string manipulation, and other system techniques. Finally, he is assumed to aspire to improve the tool as he performs his design and creative effort.

Simulation and Modeling

People who use an on-line system for simulation and modeling are classified in accordance with their role within universities. The first group consists of students who will rely on the computer system as an aid in some curriculum-related activity. They are assumed to be knowledgeable in the use of a computer and to use the system as a tool to accomplish a specific assignment. The second class contains those faculty who employ the computer to model and abstract a complex entity to derive some understanding of its functioning. The third class includes those who turn to the computer as a teaching aid in their courses. They are inventing modeling aids and teaching systems; they may, in fact, be taking the fruits of their own research and tailoring a complex model to make it useful/intelligible to the student.

Data Acquisition and Process Control

The people who use on-line computer systems for data access and process control divide according to their purposes for using them. The first class of user is made up of those who want to collect data. Often they have a real problem of capturing fleeting data. The process may range from traces in a bubble chamber on the accelerator to the number of pecks per pigeon per hour.

The second class comprises those who want to achieve automatic control of their experiments. They may also be collecting data, but on-line control is paramount. They may be delivering food to animals as rewards for responses at specific instances, sending flashes on a screen in predetermined patterns and intervals, or changing pressure levels within a hydraulic system to test the limits of its control.

The third class includes those for whom human-mediated control is important. The computer system provides information on the status of the experiment in such a way that the user can respond in a timely manner to control or react to the situation. This reaction is typically done through the computer, which then continues the experiment. Several physics and crystallography experiments are performed in this fashion.

In a fourth class of use, an experimenter has a mass of data needing prompt analysis so that he can consider what to do next. He may be relying on insight from analysis with an algorithm to establish the appropriate next set of adjustments. This form of on-line analysis is typical of modern accelerator experiments where the data from the last experiment goes to a large computer for analysis and returns to the experimenter.

Finally, there is the system that is primarily concerned with follow-up activity. Thus in a large gaming experiment the data must be collected on-line and prepared for analysis promptly in order to respond to the game's participants in an orderly fashion.

All five types of user have a set of critical requirements: (1) there must be approximate measuring devices; that is, you cannot have too remote a control system, (2) some backup control system is needed in case of failure, (3) the system typically generates large amounts of data for which an auxiliary storage device is necessary, (4) there is a need for retrieval of ancillary data; that is, a description of the patient, the time and history of the bird, and so on. Auxiliary processing is required for on-line analysis, statistical or otherwise, of the data as accumulated to indicate whether further experimentation is necessary, and (5) convenient submission to follow-up processing is needed; that is, after the experiment is over, the data are reported for total analysis.

Instructional Systems

Various groups take part in the design and implementation of instructional systems. The first and most important group includes large sectors of the student population. Typically, one thinks of these students as having a common set of needs and as being geographically spread. The instructional system is intended to justify the high cost of its development by pulling together all these needs. The second class of user consists of the educators who are developing the content or material for the course. These educators must keep in mind the students' common needs. The third group contains those particularly associated with developing the instructional program that will present the material to the students. This program will require flexible input and output mechanisms and languages tailored to the needs of the material. A fourth group of user considered by this workshop is made up of learners who are outside of the traditional educational institutions but who might be served through computer-based instructional systems on a community or home basis. Finally, the system should be designed with the school administrators in mind, since they will manage the system that provides the service.

System Requirements

Given this broad range of users, we now consider the requirements of the systems that will satisfy their needs. The workshop regarded the following requirements as the most important, and they can be considered essential for a truly operational network system. As identified earlier, the characteristics for the first user group, those interested in problem solving, may well be considered as prerequisites for the others. Most of the characteristics exist at present but are in embryonic states of development. They represent a broad range of developmental challenges from mere improved dissemination to invention of hardware.

In general, for on-line problem solving, it is essential that a computer system have the following characteristics: (1) good response time, that is, two or three seconds; (2) access by a variety of users to terminals in a comfortable and relaxed environment; (3) facilities for preparing programs off-line by writing them out or typing them in paper tape; (4) good diagnostic facilities, both automatic and requestable; (5) for all CRT applications, hard-copy options and preferably a variety of terminals; and (6) a fair accounting system.

This class of user, although small in number in relation to its potential, is probably receiving the best support in responsive computing today. The systems at Dartmouth, MIT, Stanford, and other places provide good problem-solving support with extensive documentation and reasonable response and programming accessibility. But in all these systems the terminals' flexibility is limited, very few have real backup, and most are limited in the range of activities to which a student has access for a given application.

Specific Problem-Solver Systems Needs

Each of the four levels of problem solver has a specific set of needs from a system. The canned programmer needs (1) very simple log-on procedures, (2) on-line instructions for using the terminal, (3) a large library of canned programs with an on-line index, (4) on-line instructions for program access and use, (5) on-line instructions for use during execution, including a help function when he gets lost, and (6) freedom to correct mistakes quickly and easily.

Knowledgeable users need (1) on-line instructions on how to gain access to the available facilities and languages, (2) facility with several languages, (3) very fast compilers with excellent diagnostics, and (4) a large subroutine library with an on-line index and access to information.

Sophisticated programming users need (1) access to special-purpose languages, computer systems, and packages, (2) on-line information on the processing with a method of referring a user to a human source of information where necessary, (3) adequate documentation of the particular processes the computer uses so that careful programming can save costs, (4) optimizing compilers in addition to fast, efficient compilers, and (5) access to the source code of the program library.

Ultrasophisticated users need (1) the ability to program very close to the operating system, (2) access to live operating systems to facilitate modification or change in accordance with the particular objectives they have in mind for the system, and (3) highly intelligent terminals with capability for interactive graphics and hard-copy production.

Designer Systems Needs
The designer user requires a very sophisticated system that can evolve over time. Among the special requirements of such a user are (1) a system for the manipulation of text, including functions such as word count, concordance, marginal notations, and provisions for multiple authors of documents, (2) a comprehensive file-management system, including a capability to create components, assemble and reassemble them in various ways, save intermediate versions, and recover previous states of the design, (3) complete and understandable documentation for users of design aids, probably available on-line, while interacting with the system, as well as separately, (4) access to other systems that will add to the library of design aids available locally, (5) access to other designers in a group working on a common problem, that is, sharing data files and communicating via the system, (6) a sophisticated terminal device and other components that allow interactive graphics, including high-quality image display, pointing devices, and, if feasible, color and motion, and (7) integration of design aids on different computers so that software in one machine automatically seeks data and processes on other machines as needed without user interaction.

Systems for Simulation and Modeling
All three classes of simulation and modeling users have consistent needs for special-purpose languages with features tailored to the characteristic content and methodology of their fields. To be useful, such languages must be fully documented clearly enough so that both the knowledgeable student and the involved researcher can use them. A second consistent

activity is the maintenance of files. A third is use of data analysis programs for abstracting significant parameters for developing a model and analyzing its output. Last, a flexible array of input-output options is highly desirable.

The special facilities needed by students are probably a series of canned analytical packages that can be put together quickly and a sophisticated and easily understood language to tie them together. The student's key needs are prompt turnaround and readily accessible components to plug together to develop a model. The typical student is under a great deal of time pressure. He is not going to run the model at great length, so its efficiency and effectiveness are not too important. Finally, the student needs access to statistical routines for analysis of the output.

The faculty member engaged in serious research typically will develop his model over an extended period. He therefore needs access to a stable system and the ability to modify it to meet his particular needs of abstraction. Further, the language and system should have some documentation in the program itself as, typically, a group of researchers works on large simulation models, and interaction and communication through the model and the imbedded documentation seem most appropriate.

The faculty developing instructional aids need a wide variety of test material to ensure the validity of the model and to explore the range of its output as they further develop and modify it. They also need three sets of documentation devices: one for the people to operate the model, a second for those who are going to teach with the model, and the third for the students themselves. Documentation languages imbedded in the simulation system would be most convenient for these users. The teacher needs a sophisticated system that will allow him to modify and monitor the operation of various parts of the simulation model while preventing other people from changing it. He also needs access to student files and a computer-based method of appraising student activity.

Systems for Data Acquisition and Process Control
The users of data acquisition and processing control have unusual system requirements. They draw on computers that are integrated into some processing situation, such as an experiment, a test sequence, or a natural event, and response time is critical. The criticality of response can arise from the rate at which the data are collected and the requirements for supplying feedback to the process. The five levels of action

are (1) data collection, (2) automatic control, (3) human-mediated control, (4) auxiliary analysis, and (5) follow-up activity. These can occur in combination or alone.

The real-time constraints of levels 1 and 2 are generally determined by the physical characteristics of the process, those of the last three more by the need for timeliness and by human considerations.

The most simple case of data collection is where an investigator is considering inanimate objects as subjects. These may be tones on a tuning fork or hydraulic systems. At the next level are animate subjects such as a pigeon pecking at a target or, finally, a human whose eye movements in a chess match are being observed. The task and its physical characteristics determine the data rate. Often the output characteristics and the way in which the particular set of measurements will relate to a larger body of measurements are a function of the total experimental design. An experiment on a longitudinal study of a particular set of pigeons over time presents far different storage needs than a broad sampling of many birds doing a similar task; therefore, the organization of the information accumulated in the minicomputer is also vastly different.

Target Instructional Systems
The special computer requirements for the instructional system group cover a broad range of technology. The key ingredient for effective innovation is a special-purpose language or programming system to develop, operate, and monitor the learning experience in a convenient and familiar fashion. This programming system should have a built-in capacity to adapt to a wide range of substance, from mathematics to French to chemistry. In some cases a special package tailored to the subject matter such as ACTSpeak may be required. It should have the capacity to draw on a wide range of communication aids, be they audio, color, displays, or hard copy synchronized with the lesson. Graphic capability is critical in some cases, for example, architecture, physics. The details of the language the developer uses should be crystal clear to the student. The student should have access to a control language that is understandable, straightforward, and, by and large, imbedded in the hardware of the system so that it is a mere button-punching operation. Within the overall instructional system should operate an administrative system that will allow a teacher monitoring a student's activities to identify and assist him and that accounts for how the instructional program is doing. There should be a documentation system

imbedded in both the developing and administrative systems that will allow future maintenance to be implemented with little or no effort. Finally, there is a particular need for a demonstration mode in the language, that is, one that will allow an administrator, teacher, or experimenter to demonstrate the value of the instructional set with simulated student response. A special need for users of instructional systems is information on which to base decisions on adoption of the system or programs, that is, evaluative information on the effectiveness of the system in varying educational environments.

Opportunities for Networking for On-Line Responsive Users
In general, the prime value of networking is the ability to pool the demands of a geographically dispersed set of users into one critical mass that can afford the development of adaptive and specialized software. This specific form of software support would include improved documentation and the support necessary to maintain it and make the software accessible to its dispersed clientele. In addition, the network would permit an unusually high degree of reliability by the shifting of computation centers to support the on-line use. This backup ability is unique to networking; in fact, the degree of reliability demanded by most user groups can be met only through networking.

Another form of service to special-purpose groups would be the development of data bases (and their management systems) that would be accessible to a large number of dispersed student teachers whose specialties cover a spectrum. On-line data systems seem particularly pertinent for the social sciences and may become more important as a means of cooperating with communities as time goes on. It is conceivable that the network would bring about the development at one location of a powerful special-purpose computing system that can serve nationwide needs with less effort in maintenance and less documentation of how the system works. That is, network users would require that only operational documentation be imbedded in the system without extensive system documentation and software standardization. Finally, a broad network usage might create a great demand for terminals and therefore significantly reduce the cost of each individual terminal due to the greater primary demand.

Problems Envisioned by the Network
The problems formulated in this workshop are posed as a series of questions: (1) How should services of the network be charged, and who

is eligible to obtain them? (2) What form of cash flow or credits would be necessary to establish it? (3) What is an appropriate organization for a specialized or regional network? (4) How will policy decisions be made? (5) With an option to compute locally but expensively or remotely and cheaply, how would funding policies affect the decision? (6) How can one encourage construction, documentation, maintenance, and distribution of the available sophisticated tools? (7) What about program copyright, royalties, and security? (8) How can one enforce standards for high-quality documentation? (9) How can one enforce conventions or standards for program communication? (10) What is a reasonable set of accounting systems for such a complex entity? (11) What about the financial structure and the cash flow problem? Does cash actually flow among users? At what point do institutions call a halt to paying off-campus sources for services and invest those same funds internally? (12) How do we develop and maintain a reliable data communications system, available to all, at a reasonable cost? (13) Do the economies of scale apply at all levels? Are there limitations? Is a large network best (in both economic and noneconomic services), or does there come a time when several smaller ones are better? (14) The classic question, and most often raised is: How does one cope with the usual factors of resistance to methodological and organizational change in a situation where most people have depended on their institution to provide a service and now must rely on cooperative ventures for the same service?

Opportunities for the National Science Foundation
The opportunities for experimentation in the future seem to emphasize moving from existing institutionally oriented, special-purpose systems to small nets. This would be accomplished by linking the present service systems to other sites with a designed commitment to documentation and service. This linking raises the key issues of user service and documentation, which many participants feel are the most critical problems to solve if a computer network is to function. An implicit assumption in the discussion was that there would be a hierarchy of computing power. This implies that most institutions would have some type of computer for administration, education, and research. The decision to consider what to do locally and what to do on the network is critical to planning network services. Concern for and organization of this decision should be a key element of almost all NSF experimentation.

Many participants felt, although the question was not discussed at length, that an interesting experiment would be to institute administration-supported computing in schools with no history of computing. This support might include a great deal of responsive on-line computing in scheduling, registration, and budgeting. The effort could be the beginning of a broad network of computer utilization in the institution. A project to test the feasibility and operationality could be stimulated with NSF funding to define how responsive computing can be provided to an administration by a network and how best to develop such a nucleus into a full-blown computer-supported educational program.

Software exchange would be one of the main advantages of a network. It might be feasible to determine the current availability of software by a reasonable sampling of computing centers and to experiment with a variety of procedures to share it. An experiment aimed at the dissemination of existing software and examining its usability at other places when transferred either by mail or electronically could shed light on the feasibility of a broad-based program. An interesting question is, what parameters are critical in deciding to swap program versus data? The study should consider the feasibility of sotfware exchange via a wide range of communication devices.

To focus on the computer-based design problems, it would seem practical to identify one node or source of design computing in the United States, perhaps the ICES (Integrated Civil Engineering System) at MIT on civil engineering design, and to arrange with a group of schools to utilize this system in communities where there are a large number of geographically dispersed users. A similar effort might be initiated in simulation gaming, where a number of participating institutions might use the Carnegie-Mellon Management Simulation Game.

The Office of Computing Activities of the NSF is already funding at least three projects in the "laboratory automation" area. These are at the University of Kansas (P. Gilles), the University of Chicago (R. Ashenhurst), and the University of California, San Diego (K. Wilson). The second and third of these already have a network orientation, and the first certainly has the potential. The somewhat different orientations and approaches of these projects presents the opportunity for a comparative evaluation. But the existence of three such projects also makes it possible to consider a future phase where different laboratory support groups interact with one another through networks. Since networking for systems of this type facilitates replication at other sites, a distinct possibility is the formation of a subnetwork of similar and diverse laboratory automation systems. It seems likely that the investigators in all three existing projects will be interested in exploring the possibility of future collaboration along these lines. Projects that encourage other institutions to replicate one of these systems and collaborate on developing information and documentation exchange mechanisms are also a promising avenue for promotion.

Instructional system experimentation should consider the development of systems and instructional programs. One phase is the development of pace-setting programs of high creativity and the use of networks to evaluate it in diverse settings, improve it on the basis of evaluative data, and expose it as a model for future development. The hope is that this development would stimulate studies in the implementation of instructional systems in individual institutions, consortia, states, and regions in order to provide alternate implementation models.

17

ROBERT M. HAYES
Becker and Hayes, Inc.

Bibliographic Processing and Information Retrieval

The possible applications for a National Science Computer Network in the area of bibliographic processing and information retrieval have been thoroughly explored in the literature, in a variety of experiments, and in several existing commercial services. Medline, the processing centers for census tapes, and the individual scientific information dissemination centers, all illustrate that these applications are both technically feasible and operationally useful. My comments will summarize the relevant characteristics of such applications and indicate some possibilities for utilizing computer networks.

In general, these applications are all characterized by the use of large data bases, on the order of 10^9 to 10^{11} characters. For example, Chemical Abstracts Service tapes, which are being distributed to a number of dissemination centers, now store well over 10^9 characters; census tapes, now available at a number of university campuses and commercial tape-processing centers, store between 10^9 and 10^{10} characters, depending on format. Most of these data bases have been produced as part of an ongoing program of data acquisition and publication. This is especially true of the so-called bibliographic data bases. Their producers follow their own requirements. As a result, data codes, record formats, and file organization all vary widely, and anyone using these data bases must be prepared to handle that variety.

The sources of these reference files are found in every sector of our society:

Scientific professional societies
Chemical Abstracts Service, Biological Abstracts, Engineering Index, American Institute of Physics, Psychological Abstracts, and so on.

Libraries
National Library of Medicine (Medlars and Medline), Library of Congress (MARC), National Agricultural Library (CAIN), Los Angeles County Public Library (Catalog), University of California Libraries (Union Catalog), and so on.

Governmental agencies
Office of Education (ERIC), NASA (STAR), AEC (Nuclear Science Abstracts), Census, National Technical Information Service, and so on.

Commercial publishers
John Wiley (Atlas of Mass Spectral Data), *New York Times* (Index),
Predicasts, Inc., and several others.

The users of the reference files are found in commercial organizations,
government agencies, and universities. Several universities have been
acquiring these data bases through their libraries, computing facilities,
or specially established data-base and dissemination centers. An alternate
strategy, well represented by the Medline service of the National Library
of Medicine, is to provide single-point accessibility to a given data base
through some form of computer time-sharing network. The choice between these two means of distribution is not self-evident, and each has
been adopted to some extent. Many colleges and universities are acquiring MARC tapes from the Library of Congress for use in their own
operations. The Ohio College Library Center (OCLC), on the other
hand, provides a single-point accessibility to MARC data. This highly
successful operation is servicing college and university libraries throughout Ohio and is being replicated in other states and regions.

The problems in defining how to assess the effectiveness of network
access to such data bases are many. Some of the data bases (especially
the catalogs, such as MARC or that of the University of California) are
of primary value to the librarian. They are used in selection, ordering,
and cataloging and as a support to interlibrary loan services. Others, such
as the scientific reference tapes and census tapes, are of primary value
to the individual researcher. Unfortunately, the situations of the latter
are so varied that it is almost impossible to evaluate the utility of network versus local access generally. The library context, on the other
hand, seems to be sufficiently well defined to serve as a useful basis for
experimentation.

One requirement appears especially important. The nature of interlibrary loan requires access not only to catalogs but to the books and
journals that they reference. Since it may provide a paradigm for network development, it is worthwhile to examine interlibrary loan and
the role computer networks can play in it.

Interlibrary loan is the process by which libraries share their resources.
To date it has generally provided that a library can borrow material
from another library on behalf of its own constituency and will re-
ciprocate by lending its own materials to other libraries. No charges
are involved except for photocopy or other special services. The interlibrary loan process has been formalized in a standard procedure,

the National Interlibrary Loan Code. First promulgated in 1917, it has been revised several times, most recently in 1968. The basic concept of the National Interlibrary Loan Code is to reduce the burden on the lending library as much as possible. The borrowing library makes sure that the material is properly identified and that the lending library has it before transmitting the request.

To identify the lending institution is therefore a fundamental requirement in interlibrary loan. Satisfying it means that the borrowing institution, or a bibliographical center serving it, must have access to union catalogs and lists of holdings. Storage of these on-line, with network access to them, could therefore represent a service of a National Science Computer Network. Some developments, such as OCLC, are consistent with that approach. Unfortunately, the size of catalog data bases in comparison with their frequency of use raises some serious economic problems, especially early in the development and experimentation with networks.

Yet there is a related requirement that is necessary for achieving on-line catalog access but that has value in its own right and is significantly simpler to implement. It is the requirement for message switching, referral, and accounting for interlibrary loan communications. The need for this capability arises not only from the inherent nature of interlibrary loan—that is, the fact that it requires communication between institutions and frequently additional referral from one to another—but from the economic fact that the major lending institutions can no longer afford to provide services without compensation for their costs. Some means must therefore be provided for accounting for their services as the basis for reimbursement. Specifically, a recent study undertaken for the Association of Research Libraries, which represents the major lenders in interlibrary loan, indicated that the lending institution incurs a cost of nearly five dollars when it fills a request. Since a typical large library may fill as many as 100,000 requests a year, the magnitude of its costs cannot be ignored.

As a service of a computer network, then, a system for communicating interlibrary loan requests and accounting for them has considerable value. But if the system includes the capability for referral, it will also provide the means for such referral to automated, on-line catalogs to which the network may have access. The system for interlibrary communication would therefore represent a natural and necessary step toward the larger aim of bibliographic access.

The concept of the system for interlibrary communication is that interlibrary loan requests would be transmitted by input from a terminal (a teletype, for example), received by the computer network, and sorted and forwarded by it to the destination defined in the request. This message-switching function is an integral part of the system software of virtually every national time-sharing system today. Where the message did not identify a destination, where the communication protocols require intermediate destinations (such as approval points), or where the message specifies alternative destinations, the message can be automatically referred to bibliographic centers, on-line catalog data bases, referral centers, and alternative lending institutions.

Finally, since the messages identify the institutions as well as the request, it is possible to provide automatic logging and analysis of traffic, automatic accounting and billing, automatic checking of validity, and other services of value to both the borrowing and lending institutions.

This system would seem to be one especially appropriate for experimental use on a National Science Computer Network. Most of the major lenders are university libraries; a large proportion of the lending involves scientific journals; the development of library networks has been predicated on access to major university collections as backup resources; the principle that services should be recorded and compensated is already a part of several network contracts.

To summarize: the area of bibliographic processing and information retrieval, represented by on-line catalogs and reference data bases, is one already operational on computer networks in several specific applications, although most are limited in size and scope. It will become one of the most important areas of use of a National Science Computer Network. A possible area of experimentation is in the use of the network for message switching, referral, and accounting for interlibrary loan requests.

18

Text Processing and Information Retrieval

Report of Workshop 4

WILLIAM F. MASSY
Stanford University
Faculty Discussion Leader

Contributors:

FRANZ ALT
American Institute of Physics

ROBERT G. CHENHALL
University of Arkansas

DENNIS FIFE
National Bureau of Standards

ROBERT M. HAYES
Becker and Hayes, Inc.

JACK HELLER
SUNY at Stony Brook

DONALD HILLMAN
Lehigh University

JAMES F. MELLO
Smithsonian Institution

EDWIN B. PARKER
Stanford University

DAVID PENNIMAN
Battelle Memorial Institute

BEN R. SCHNEIDER, JR.
Lawrence University

VIRGINIA STERBA
New York University

BENJAMIN SUCHOFF
SUNY at Stony Brook

RONALD WIGINGTON
Chemical Abstracts Service

Introduction

This workshop was concerned with systems designed to capture, store, and selectively retrieve textual information and data bases requiring large-file capability. The emphasis was on users of systems that incorporate text processing, indexing, and searching operations. For example, systems that provide a reference-citation service, like Chemical Abstracts Service, and that allow the user to build as well as retrieve from files were discussed.

The workshop attempted to address the possibilities and problems of networks for users of text-processing and information retrieval services by means of three questions:

1. How can users be classified? What are the requirements of each class? What kinds of user groups would networking benefit most? What systems characteristics are important for meeting the needs of potential users?
2. What critical problems need to be overcome in order for any kind of network program to be successful? What research and development are needed to achieve solutions?
3. What should be the main objectives of the National Science Network research program? How should the development of a possible trial network be approached?

Edited versions of the reports of the subgroups that worked on each question and some related material discussed in the workshop follow.

User Characteristics

Classification by Professional Characteristics

Two major classes of users seemed appropriate for identification: academic and nonacademic. The academic community was considered to be of immediate if not primary concern. This class consists of

- The scientist and his humanist counterpart, the scholar, both involved in construction as well as application of data bases.
- The student, whose main interests probably will be in the areas of bibliographic files and computer-aided instruction.
- The administrator, who will need facilities for processing student records, admissions office procedures, and similar tasks.

The nonacademic user class includes government personnel, charitable organizations, independent research groups and individuals, the business community, professional societies and their members, and, indirectly, the general public. This class will need access to large data bases such as legal and medical files, census data, and other general or special compendia. It is possible to imagine that the household now equipped with a telephone will someday have a terminal as well; thus the doctor, housewife, or homebound invalid may have an impact on networking.

Classification by Type of Data

The needs of users derive not only from their occupational characteristics but also from the kinds of data they normally search. By far the greatest and most universal need is for bibliographical information. Bibliographies and indexes may be general (a library catalog) or highly specialized (a personal file).

Next in order of importance might come other reference works basic to the various disciplines, such as dictionaries, special catalogs, annuals, archives, public records, registers, concordances, personal catalogs, tables, and so forth. These data may be verbal, like parish registers, or numerical, like census or voting data. Another important need is for information about collections of objects such as museum catalogs, archeological collections, and inventories.

These kinds of data come in lists having a reiterated structure. Some users may also require actual text resulting from a bibliographical search, or a visual or auditory representation of an item. The text of a literary work or historical document may be required for analysis of content or style.

There will also be a need for compilations of observational data that may or may not have been subjected to some prior form of processing. These data include such things as weather reports, results of behavioral experiments, measurements, surveys, and the like.

Classification by Type of Use

No taxonomy of user characteristics could be complete without consideration of type of application. Two general categories were discussed.

1. DISCIPLINE-ORIENTED APPLICATIONS. Here the user's needs for information derive from the nature of the subject mater and the traditions and standards of the field. More specifically, such systems are intended to help individual researchers and scholars to teach or advance the state of knowledge of their disciplines. Examples include general and special bibliographies, classification-oriented data banks in the natural sciences, and thesauri. Data structures and search procedures must fit the needs of the discipline. Response time and other system parameters are geared to the individual user—scientist or scholar.

2. PROBLEM-ORIENTED APPLICATIONS. These systems are designed to help solve operational or managerial problems. Examples are vital organ registries, tissue typing, and patent records in medicine; accounting, information, and other administrative systems; and library automation (for example, acquisitions search and technical processing). Data structures and system parameters derive from the nature of the problem, often expressed as an explicit system specification.

The group felt that both types of applications should be studied as part of any National Science Network research program.

Classification by System Configuration

At least three types of system configuration can be identified. They are related to the degree of coordination and control available or desired at each of several stages in the acquisition-storage-use cycle.

1. DATA DISSEMINATION SYSTEMS. Here a single operating group captures and processes data into a file, which is then made available to a larger and perhaps quite diffuse user group. The file may be maintained centrally and reached by teleprocessing, or various "slave files" may be maintained or copies made. However, all slave files are effectively "read only," in the sense that users normally do not make contributions or updates without going through the central authority. Examples are the MARC service of the Library of Congress, Chemical Abstracts Service, and the U.S. Census.

2. DATA-SHARING SYSTEMS. Multiple operating groups or individual users contribute to and use data in the file. Since contributions are intended to be generally available, one file must be designated as "master," though copies or slaves are possible if periodic (batch) update is acceptable. Examples are the union catalog feature of the Ohio College Library Network (which itself makes use of an external dissemination system—MARC), and various museum and science information exchange services.

3. "PERSONALISTIC" INFORMATION STORAGE AND RETRIEVAL SYSTEMS. Here the individual user is given the capability to use generalized system software to design his own files, insert and maintain his own data base, and do his own search-and-report generation. Sharing with other users is possible but not mandatory. An example is the Stanford Public Information Retrieval System (SPIRES).

Table 18.1 summarizes the foregoing three-way classification by system configuration.

Table 18.1. Three-Way Classification by System Configuration

	Data Acquisition	Files	Users
Data dissemination	Central*	Central*	Diffused
Data-sharing	Diffused	Central*	Diffused
Personalistic	Diffused	Diffused	Diffused

*"Central" refers to a server unit, not necessarily to the central network management.

These three configurations may require different network organization and policies.

Classification by User's Skill Level
It is obvious that the potential users of any network will span a wide range of skill levels vis-à-vis computers and related systems. The workshop members agreed that it is desirable to consider all levels of potential users of networks. One of our later recommendations speaks to the issue of early priorities. Here we merely categorize the several levels of skill for the record.

1. THE "COMPUTER-MATURE" USER. He has developed sufficient skills to use whatever system he is addressing with a fairly high level of competence. Generally, however, he is not a computer scientist.

2. THE "CAN DO" USER. This person is not well trained in the relevant application but is willing and able to learn. This class represents a large potential, but it is also a possible drain on the user education and consulting resources of the network.

3. THE "WHAT CAN IT DO FOR ME?" USER. This person or group may have a significant application but is unlikely to put forth much effort to define it in computer terms or to learn to use even a moderately complex system. Getting this kind of user involved will require a very high degree of "user orientation," including perhaps a specially tailored system and sometimes a concerted educational effort. An example of a noncommitted user who might nevertheless be of considerable importance to a network is the practitioner of clinical medicine. Some potential users of management information systems may also fall into this category.

The "target market" for network services must be defined in terms of these or similar categories of skills. Developing this market may be done by participating institutions, consortia, or the network management itself.

System Characteristics

Characteristics for the "Minimum-level" User
We have just classified users according to their level of skill. In addition, a user's institution (if any) may approach the system as either (1) a participant in the process of data-base construction, (2) a cooperator in data-base or system maintenance, (3) a "server" or provider of the system, or (4) a user of services and data. The lowest common denominator for both categorizations is the person with a lower skill level at an institution who has no involvement with the system except as a data user. The system must cater to this type of user by being easily approached, simple enough in outward appearance to be comprehensible to the non-expert, and sufficiently sophisticated and responsive to meet the needs of all classes of users.

The following specific capabilities and services have been identified as necessary to minimum-level users.

1. Easy access to a data base through terminals. A querying capability that is interactive, expressed in a fairly natural language, and reasonably forgiving of mistakes must also be provided.
2. File control procedures that permit easy access to authorized persons and govern the data that can be obtained by people with various levels of authorization. These procedures must be fairly transparent to the average user.
3. Extra computer communication to facilitate interaction among users and transmission of information for prospective users. Particularly urgent is the need to develop data standards for the treatment of bibliographic and other types of textual data.

Many users of bibliographic and other textual data do not currently demand real-time response. The means to receive this type of response have never been available before, and considerable adjustment in how data are recorded and how users frame their questions will have to come about in response to this new information-handling power. Nevertheless, considering the great volume of textual data already in existence and the continued acceleration of its production, there can be little doubt that it can be controlled only through the use of computers. Penetration of the market for bibliographic applications will depend heavily on the availability of interactive systems for framing questions and obtaining at least preliminary results.

General Services to Users
The subgroup working on this question identified the following as

"necessary" or highly desirable services that a network should provide. Here the discussion is not limited to the minimum-level user. The services are listed in approximate order of priority.

1. The user must have access to comprehensive documentation of system facilities, capabilities, and software geared to *all* levels of user sophistication and encompassing an on-line HELP system. This feature must be combined with consultation services for those seeking advice. Both the producer and the consumer of network resources must be considered.

2. Retrieval capabilities, preferably using natural-language input, should be offered at different levels of response time from immediate (on-line, in seconds) to hours (batch mode of operation) or even days (retrospective searches done inexpensively on a time-available basis, with output mailed).

3. General-purpose software packages for data analysis of large files are a requirement, both for statistical studies desired by the scientist and for literary analysis needed by the scholar. This latter category includes content and stylistic analysis, concordances, and so on.

4. The knowledgeable user (organization or individual) with a "private" data base should have access to file management software so that he can "massage" and update his own information. Such programs include text editing, support for the creation and maintenance of files, and housekeeping in general.

5. Users should have available to them data bases in related fields. For instance, a student researching eighteenth-century theater may be interested in searching files in eighteenth-century music as well.

The following "desirable" features are also listed in order of priority:
1. A network could encourage the development of standardized representations for data concerning nonverbal material like music and art. Where this is impossible, suggestions for translation procedures for incompatible data might be feasible. The point is that the force of network opinion might help make a given existing system acceptable even to advocates of other styles.
2. The option for "high-quality publishable"—that is, printout would offer enormous benefits in terms of time and cost of reproduction, which is a totally redundant step.

It should be clear that a large-scale network system can economically provide many of the services just listed where small demand at a local facility may make user satisfaction impossible. For instance, one installation on the network could purchase an extended print train with lower case and special symbols for esoteric scripts such as foreign languages or mathematical formulas and make it available to the entire system.

Critical Problems

The subgroup on this topic considered nine major problem areas in the development of networking. While the basic orientation is that of the user of text-processing and information-retrieval systems, the discussion has obvious applicability to users of other kinds of systems. The areas considered are (1) bringing institutions and services onto a network; (2) communication and hardware components; (3) steps in the implementation of a network; (4) overlap, compatibility, and standardization; (5) problems relating to data base management; (6) security and privacy; (7) data evaluation and quality control; (8) the interface between network and user, and (9) public access. These issues and suggestions for study are offered as a contribution to the agenda of the National Science Network program of research.

Bringing Institutions and Services into a Network

There are at least two strategies for bringing institutions into a network: as groups, in consortia, with services for them defined as part of a contractual arrangement; or individually, with any or all of the services available to them as an integral part of their membership.

The former strategy seems to offer several advantages. It allows existing arrangements to be integrated into the network while preserved in their existing structure, financial and otherwise. It also allows competition among institutions to be resolved in the contract process rather than in the actual operation of the network. It provides a means of gradual expansion of membership and services in a step-by-step progression. Finally, it provides a means of protection of network resources from unwanted overload, since the contractual arrangements would specify conditions of their use.

Communication and Hardware Components

Again, there are at least two strategies: the network could include development and management of the hardware and communications components, or it could limit itself to managing the use of those components and to the negotiation of agreements with institutions or consortia.

Here latter strategy seems to offer several advantages. It allows the network to choose among alternative communication and hardware systems without committing its own capital. It allows the management talent of the network to be concentrated on the development of usage and on relationships with the user community. And it allows the network to represent better the interests and demands of the using community in negotiation with the communication and hardware contractors.

Steps in the Implementation of a Network
Those who design and develop the network will wish to be guided by pilot studies or experiments on a relatively small scale. But such efforts are impeded by the fact that the small study may not allow scaling up. For instance, an experiment in indexing, based on a few thousand items and a thesaurus of a few hundred index terms, does not even begin to reveal the difficulties encountered with a million items and 50,000 index terms. For another example, using a narrow field like thermodynamics for a pilot study may be quite misleading when the results are applied to, say, nuclear science or electronics. Thus in many cases developers would need large-scale experiments, which involve prohibitive cost.

An alternative that seems to be available in the NSN Program is to use existing subsystems, either unchanged or with minimal changes. Possible candidates are regional or specialized networks and computerized data bases. This approach has the added advantage that the results are useful in their own right even before the entire system has been created. It involves the risk of slanting the outcome in the direction of those systems that happen to be conveniently available.

Overlap of Coverage
Since the network is to serve a number of different purposes, it will have to offer a number of different data bases, and it seems unavoidable that there should be some duplication among these. For example, suppose that the network makes available bibliographic data bases in different fields like philosophy, mathematics, physics, chemistry, biology, and medicine and that these data bases are prepared by different organizations. Clearly, portions of the literature will appear in two or more of these data bases. While this leads to added cost and some confusion for the user, it appears to be an unavoidable condition at this stage of development.

Problems Relating to Data Base Management
Various policy and administrative problems as well as many technical problems are associated with data-base management in a network. The technical problems of file organization, access routes and methods,

backup and recovery, software efficiency, and the like are associated with the mechanism of how the functions can be carried out. The policy and administrative problems involve what is to be done by whom under what conditions.

The network as a marketplace must accommodate the parties involved—information sources, information users, and user assistance services—in solving these problems. The following problems are basically the same whether one is considering original text, access to bibliographic and subject indexes or retrieval and use of numerical data. All are referred to here as "information."

1. The sources and the users must operate on compatible bases of definition of information elements, structures, and representations. Especially where information is to be coordinated and merged from multiple sources, the common items must be handled compatibly. The mechanism for establishing and updating these definitions must exist.

2. Where information storage services are provided for user-generated or user-compiled information, the standards and limits of responsibility for safekeeping are important. Similarly, reliability of delivery of information from any source is important.

3. Accounting mechanisms must exist to facilitate the determination and collection of revenue to support the building and maintenance of information collections.

4. A specific policy to encourage research and development to solve the many problems in the management of very large files is needed.

One cannot expect the network itself to cope with all problems of compatibility. Standardization is the concern of several other organizations that have been on the job for years and are likely to continue for years more. In fact, new incompatibilities probably arise faster than the old ones are resolved. Furthermore, data bases coming from different sources are designed for purposes other than their use in the network, and these other purposes frequently dictate formats that are incompatible and that often defy automatic conversion. This is a subject on which considerable attention and research need to be focused.

Security and Privacy
The problem of security and privacy is a specific subset of the questions of data-base management and the transmission of information. Preservation of the integrity of information and establishment of ways to guard access to specific information according to source-user agreements will be

vital. In addition, statistics on the network's own traffic as they associate specific users with specific topics may be sensitive information. Concern may arise in the case of commercial organizations or other competitive situations. Policies with respect to the form and availability of such information must be established and enforced.

Data Evaluation and Quality Control
The cost of certifying the quality of data in a data bank is considerable, whether the data are bibliographic, public opinion polls, measurements of physical properties, or something else. A program of data evaluation conducted by a network could lead to high costs and to political difficulties with the groups whose data are being evaluated. Yet failure to distinguish quality-controlled data banks from haphazard collections may also lead to problems. Being merely a "common carrier" service with a variety of data banks and services made available on a *caveat emptor* basis may be a disservice to users who come to the network hoping for assistance in choosing a data bank or service that fits their particular needs.

A third course might be preferred. This is to provide a *Consumer Reports* kind of service describing the quality of different data banks and services. Such a service would be a useful adjunct to a switching and referral service but would not prevent any data bank from being made available nor greatly increase the costs of the basic services offered.

Interface between Network and User: Research Needed
Present computer systems and networks require a nonprogrammer user to undergo a certain amount of training or initiation before he can comfortably use them. Two factors will tend to make this problem much worse as networks develop. One is that as the number of users expands, the percentage of those who are sophisticated in the use of computers and tolerant of computer foibles will decrease. Early users of data files may have computer backgrounds or may be the type who will, readily adjust to the system's ways. Later users are likely to be more critical and to expect a smoothly running service that somehow understands and adapts to their needs. The second factor is that the number of users at remote locations will increase, greatly compounding the problems of providing training, consultation, and user manuals.

Satisfactory solutions to these problems are not immediately available. Research is needed to develop and improve on-line tutorial and consulting techniques, and documentation, self-teaching systems, and general ways to make the interface of systems and users more adaptive to the needs of relatively inexperienced people.

Public Access

If we assume that the network functions primarily as a switching center, referral service, and accounting service for a variety of decentralized services, the question of who is permitted to have access to any particular service remains a policy question for each server node on the network. For example, some nodes may be operated as commercial businesses with service available to any member of the public who wishes to purchase from them. Other nodes, because of nonprofit or tax-exempt status or because of educational discounts obtained from hardware manufacturers, may be prevented from providing general service. Still other nodes, as a matter of policy, may choose to serve restricted classes of users such as students and faculty in institutions with whom an appropriate institutional agreement has been made.

The network itself should add any restrictions on the criteria for selection of users other than those imposed by the server nodes themselves.

The consensus of this workshop was that any national network involving colleges and universities should grow from existing user communities concerned with storage and retrieval of information dealing with text or data and from special computational facilities. Participants would represent an established regionally or nationally recognized user constituency or would otherwise have participated in some resource-sharing arrangement. (This suggestion is intended to assure maximum probability of success by building from a foundation of existing capabilities. It is not intended to restrict access to the network by any individual user. This would be mediated by the user groups and not the network.) A fundamental aspect of the information-transfer activity would be to foster the development of special-purpose, tailored information products, services, and systems.

In the short run, an appropriate organization might undertake the following tasks related to networking:

- Continually monitor the special needs of user consortia and establish mechanisms to meet them.
- Maintain a catalog of programs and data bases and use it to provide a referral service.
- Conduct an active educational and promotional effort intended to

bring potential participants up to a level of expertise before entering the network.

By this approach the trial network would be able to supply a needed service for participating institutions while at the same time demonstrating the feasibility of networking in a college and university environment. Investments in hardware and systems development would be minimal. Significant resources can be devoted to the key problem areas of making a cohesive market for existing computer services at scattered locations and providing the user education needed to apply these services to important scientific and pedagogical problems.

III Organizational Matters

Part III contains the papers and workshop reports of the seminar on organizational and policy issues. Participants ask such questions as
• What are the operational characteristics of organizations now sharing computer resources and providing broad services? What changes seem necessary to inspire cooperation and coalesce users and suppliers? How are new viable organizations started? How does one gain the support of users for a long-term commitment? How are changes effected in institutional attitudes to share and receive computer resources? Should existing organizations review their goals and objectives relative to opportunities for growth, change of venue, or controlled reduction of services?
• Once a critical mass of a shared network service is achieved, how is the network organized and managed? What are the appropriate time dimensions for the establishment and maintenance of a resource-sharing network? How is it operated? What are the critical operational issues?

There are nine general background papers. George Feeney presents the management and organizational issues faced in establishing a concentrated worldwide commercial computer network and describes the physical characteristics of this network (Chapter 19). James Emery discusses the systems implications of computer resource network sharing (Chapter 20). Charles Warden analyzes the economics of sharing (Chapter 21). Zohrab Kaprielian considers the politics of cooperation (Chapter 22). Chadwick Haberstroh examines the individual behavioral implications of organizational change (Chapter 23).

The remaining four papers report on ongoing operations and provide a springboard for discussion in the four workshops (5–8). Leland Williams presents the Triangle Universities Computing Center network management system (Chapter 24). Robert Gillespie examines the role of the University of Washington Computing Center in the Western Washington Network (Chapter 26). Harrison Shull discusses resource sharing in theoretical chemistry (Chapter 28). Thomas Kurtz outlines the evolution of the New England Regional Computer Network (Chapter 30).

The workshops consider existing organizational structures and the particularly important problems that must be faced in the development of expanded networks of computer systems, including the economic aspects, the political implications, and the management tasks, as well as the organizational issues involved in establishing and maintaining an ongoing network of computer resource sharing. They also consider how present organizations can evolve to meet future needs and what the role and form of these organizations should be.

Workshop 5 examines the technical, logistic, communication, and management issues that have to be faced to organize and maintain over time a service for decentralized users and suppliers (Chapter 25). Organizing and running a computer network involving different institutions, decentralized users, separate affiliations, and dissimilar interests presents special management problems. Workshop 5 takes up these problems, looking at the relative advantages and disadvantages of alternate approaches to managing the network, identifying the new organizational entities that might help, and discussing how to get started building for the long term.

Workshop 6 explores ways of organizing and managing a university computing center that best utilize a network of computer resources (Chapter 27). It examines the possible conflicts with internal university policies that might arise in establishing a network relationship, recognizing that an interuniversity network may offer a university services that it would be difficult for it to provide itself but

may also raise difficult questions of an organizational and political nature. The university computing center may have to redefine its role within the broadened context of a network, and the university administration may have to find new collaborative ways to relate to its sister institutions.

Workshop 7 considers the characteristics of existing user organizations and asks how users should be organized to maintain effective communication with suppliers and to reflect user need and resources (Chapter 29). The most effective way to locate, share, and distribute resources in one field or discipline may or may not be the most appropriate way to do it in another. Users can organize themselves either by professional interest, mode of computer use, or supplier with whom they associate. Workshop 7 explores alternative ways of establishing joint computer facilities and forming user groups.

Workshop 8 looks at the organizational and political issues that regional users must resolve to maintain a viable service and discusses the boundary conditions and entry commitments that seem reasonable for regional networks to adopt (Chapter 31). In the course of examining the possible organizational relationships between regional and national networking, the workshop asks whether a national network could be formed by welding together a number of regional networks. One issue is the need to bridge state and other jurisdictional boundaries. Another is the degree to which the system runs the user or the user runs the system. Small institutions do not want to be swallowed up, and allocation mechanisms are needed to make sure that no user exerts undue influence, no matter where it is located or how large it is. This is one of the several problems for small users that is discussed.

19

GEORGE J. FEENEY
General Electric Company

Concentration in
Network Operations

First, I would like to describe our network at General Electric, both because it is interesting and because it illustrates the issues of policy and organization transcending technical matters that are the subject of this discussion. Second, I would like to express, as an outsider and former user of computing resources in universities, what I consider to be some central issues in the evolution of computing in higher education.

The General Electric Network

Figure 19.1 shows the present coverage and major features of the General Electric network. The system provides low-speed access to large central computers located primarily in Cleveland from remote sites throughout the United States, major portions of Canada, Alaska, and most of the principal countries of Western Europe. Users, whether in Milan, London, Anchorage, or San Diego, simply make a local telephone call and thereby gain access through a low-speed terminal to a system in Cleveland. Typically, usage begins at 8 A.M. in Rome and ends at midnight on the West Coast, which just happens also to be 8 A.M. in Rome. Thus we are providing first-shift service twenty-four hours a day. This requires a complete rethinking of how a computer center should be operated and of what commitments it must make to its users.

Our probable plans for the next twelve months include an extension of the network to Japan through a Japanese distributor (pending final authorization from Japan); a similar extension to Australia (assuming the market there is sufficient to justify the cost of the satellite channel); and extensions to Mexico, Brazil, and a number of the European countries not yet on the network, including Switzeralnd, Denmark, and Norway. The network is unique in its geographic coverage and is being actively used.

There are about 2,000 access ports in the United States and more than 100 additional ports in Europe. These ports are deployed through regional distribution networks that include about 80,000 line-miles of transmission, some of which is multiplexed by means of conventional frequency-division multiplexors. The regional networks converge into sixteen distribution points—staging points for store-and-forward transmission—in such cities as Los Angeles, Berkeley, Dallas, Atlanta, London,

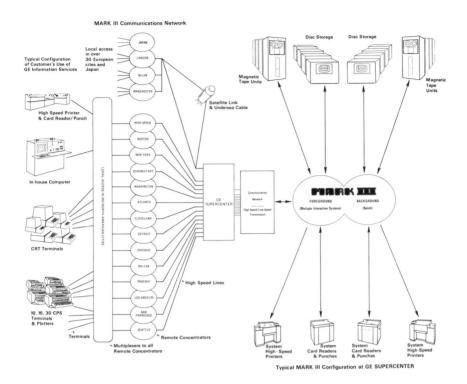

Figure 19.1. MARK III communications network.

and Milan. The sixteen distribution points have stored-program computers that serve as remote concentrators. They correspond to the IMPs
in the ARPANET. The remote concentrators do not talk to one another,
and consequently traffic cannot be moved directly between them. Each
remote concentrator has up to ninety-six low-speed lines emanating from
it and two high-speed lines, operating at either 2,400 or 9,600 bits per
second, which collect and channel traffic back to a central concentrator.
Service continues normally even if either one of the high-speed lines fails
since each line can handle full traffic. Because of this extra capacity, the
system is able to suffer overload from time to time without serious effect
to the users. Since the transmission discipline is store-and-forward, line
noise occurring between the remote concentrator and the central system
can trigger retransmission. This provides for essentially error-free communication.

The transmission lines from the remote concentrators are gathered in Cleveland into central concentrators, each of which can normally handle up to eight remote concentrators. The central concentrators, General Electric GEPAC 4020's, are large-scale stored-program machines. They were originally intended as process computers and have very high reliability. Two additional 4020's serve as switching computers that interconnect the central concentrators and by this means supply closure to the network. That is, through the switching computers a path can be established from any access point to any central system. Each central concentrator can provide direct communication access to one or two central systems. The central systems in Cleveland consist of Honeywell 635 and Honeywell 6000 processors. They are backed up by IBM and CDC mass storage devices with a total capacity of several billion characters.

Figure 19.2 gives another look at the network. This mosaiclike diagram is a stylized view of the network as seen by the people operating it. The outermost ring represents the remote concentrators—Los Angeles, Schenectady, Chicago, and so on. The lines connecting them to the next inner ring are land lines going from the remote concentrators into the central concentrators (in Cleveland and one in London), which are connected to each other through the switching machines. The next ring shows the main frame systems with their permanent storage. In practice, the systems are designated by letters that serve as routing characters. When a user logs in, the first character of his identification number represents the central system, such as N, that he will be using.

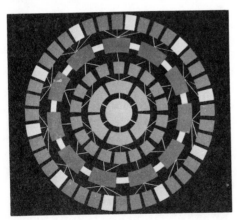

Figure 19.2. Stylized view of the General Electric network.

The innermost ring designates the remote-batch systems, which are Honeywell 6000 computers with 262,000 words of core memory and a full complement of tape handlers. They use the GCOS operating system. The remote-batch systems are connected to foreground time-sharing systems by wide-band links that allow a time-sharing system to be used to prepare work for a remote-batch processor. Even more elaborate combined operations are possible on the system.

Within the coming year, we are planning to add IBM main frames to background mode. This will provide an alternative form of background computation that will be compatible with the majority of customer systems. We are considering the possibility of putting CDC or Univac large-scale systems in background as well. Such systems in background, available instantly to half the world, would open up some interesting possibilities for a certain class of large-scale scientific computations.

Among the array of remote concentrators in Figure 19.1 are shown several high-speed concentrators that provide service to terminals such as 2780's and Data 100's operating at speeds of 2,000, 2,400, or 4,800 bps. Thus, high-speed terminal access is also available through the network, although low-speed access is still the dominant mode of use.

The key point about the network is that it is a system of nearly 100 computers, each of which in its previous life was a free-standing stored-program machine that operated independently of other machines. The service derived from the network is obtained by forcing these machines to work together. By their design they are somewhat reluctant to work together, and they are interconnected through an equally reluctant communications medium consisting of voice-grade telephone land lines and derived voice channels on transoceanic cables and satellites. Some of the means developed to overcome these difficulties have already been described elsewhere.[1]

Marketing and Support

There are in total 1,000 people associated with the operation, marketing, and support of the network. Their productivity in terms of revenue received per employee is about twice the average in the General Electric Company. Of the 1,000 people, nearly half are in the field in sales and support roles working with and directly available to customers. The remainder are concerned with maintaining the network and running the machines, developing new software, planning, and just trying to make it all work as one system. As an indication of the complexity of the job, we get approximately 1,500 telephone bills a month.

Although we have a large field sales organization, sales is probably the wrong name, since about 75 percent of their energy is devoted to *support* of present customers: training, answering questions, helping people with manuals, and so on. Sales in the sense of knocking on doors and trying to get someone to sign a contract occupies only a small percent of the effort of the group.

Support is really critical. It is the requirement that is most likely to be underestimated. It takes an enormous amount of support to make the system useful to the end user.

Concentration of Computing

We made an exciting discovery in 1968 that we could hardly believe at the time, although it seems quite obvious now. Based on some fairly simple assumptions and some data on the diurnal characteristics of our load, we discovered that we had a technology in hand (developed almost through a serendipitous process) that made it possible to deliver nationwide or even international computer service at a cost actually lower than that of local service because of the opportunity it afforded to diffuse peak usage. This is the store-and-forward technology of remote and central concentrators.

A fundamental characteristic of commercial interactive computer use is that it reflects people's normal working habits. Activity begins when people come to work in the morning, picks up until about 11:30 A.M. when they go to lunch, drops off until about 1 P.M., starts back up and hits a peak at about 3:45 P.M., and then starts declining slowly until it reaches a negligible level between 8 and 10 P.M. This means that the capacity of a system is determined by what happens at 3:45 in the afternoon, unless one is satisfied with degraded service or unless one tries to change people's work habits, which is very hard to do.

Because of the time difference, the East Coast high coincides with lunchtime on the West Coast, which is just marvelous. The cost of transmission to the West Coast is considerably less than the savings achieved through better utilization of the central system capacity. We can provide time-sharing service to California remotely from a central system in Cleveland at less than it would cost from a local system in California.

In addition, a very significant part of our market consists of multi-location network applications: for example, multilocation companies doing order entry, inventory control, production scheduling, and so on. These applications make most sense on a centralized network rather than a system of dispersed computers.

There is a third reason for our preferring the centralized approach. It is almost impossible to manage multiple centers. Back in the 1960s we had fifteen completely free-standing centers in the places where we now have distribution points. At any given time at least half of them were performing unsatisfactorily. Even though there was theoretically only one service and one headquarters, there were in fact surprising variations in procedure and in the delivered product. That is tolerable in the early stage when the product is simple and the business highly entrepreneurial. But beyond a certain scale of the business, management of multiple centers becomes a nightmare.

Concentrated operation afforded us substantial leverage. It is easier in one location to attract the very best people. In addition, by centralizing we were able to invest in a protected power supply that cost nearly $1 million. It helps to protect all the main frames and other electronic gear in the center from the uncertainties of the electrical environment. We use electric utility power, but only to recharge a very large set of batteries that drive the computers through a solid state inverter that converts the battery power to alternating current. With all transients, spikes, and effects of electrical storms removed, we are able to run at better than 99 percent availability. We could not do it without this special power supply.

Thus, concentration gives leverage in the management and operation of the system. It also gives leverage in the increased levels of reliability and security it makes possible. We deliberately selected a technological approach that we hoped would give us access to new markets, such as networking, and at the same time enhance our position in our present time-sharing market. We now receive a substantial portion of our business from networking, that is, from applications giving multiple-location access to a common set of files and programs. This is business that we would not otherwise have if we were not concentrated in a manner that allows us to do networking. Nor would we have as much as we do of our base business of local time sharing, which makes up more than one-half of the total, were it not for the reliability and security that we can demonstrate because of the way we operate.

When we made the decision to go from fifteen centers to three and then to one, one of the main objections concerned the absence of local presence. The salesmen, we were told, would not have the same emotional drive that they did knowing the machine was right there in Wellesley, Massachusetts. There were people who said it would never be the same; that there would be morale problems if we took the machine out of

Wellesley or Atlanta or Berkeley. It was a tough decision with many skeptics saying the customers would not like it and everything would collapse. We had much anxiety in making that decision, although we did not feel at the time we had any choice.

Nothing like what was predicted happened at all. The salesmen did not care where the main frame was, only what kind of service was delivered. I think there is a lesson here for the university who fears that its whole computer science department will leave if the main frame is not right down the hall.

Our remote domestic users do not pay anything additional for communications no matter where in the United States they are. We can, in fact, demonstrate that the per user cost is actually lower, and that users help balance system loads by observing their normal business-hour utilization patterns within their own time zones.

Conclusion

With respect to the question of centralization, my feeling is that it no longer is a question. Data processing is going to be centralized. The only question is whether an institution is going to resist centralization—at its peril—or make centralization work for it. I am convinced that the free-standing university computer is soon to become an anachronism, about as relevant and common as a university power-generation facility, of which few still remain. Universities were once on the leading edge in the development of power theory. It was necessary to have power-generating facilities in order to carry on the academic work. My guess is that those institutions that today are the intellectual leaders in electrical engineering were the first to give up operating their own generating facilities. And allowing for some possible exceptions at Dartmouth College and other institutions engaged in educational networking, I believe the same thing will happen with computers.

Reference
1. George J. Feeney, "The Future of Computer Utilities," *Proceedings of ICCC, 1972.* The First International Conference on Computer Communications, Washington, D.C., October 1972, pp. 237–239.

20

JAMES C. EMERY*
University of Pennsylvania

Problems and Promises of Regional Computer Sharing

Origin of Regional Computing in the Delaware Valley

Like educational institutions in several other geographical areas, a number of colleges and universities in the Delaware Valley have banded together to form a regional computer facility. The University of Pennsylvania now obtains much of its computing services from a not-for-profit organization named UNI-COLL. After nearly a year's experience we are now in a good position to assess the problems and promises of regional computing.

UNI-COLL was formed in the summer of 1972 from the computer center of the University of Pennsylvania. Before this the University had pursued a conventional approach to computing. Most academic computing was obtained from a University-run computing center. A separate center was operated for administrative data processing, and several research facilities had their own computer. We had gone through several successive upgradings of machines. Each new machine brought a significant increase in capacity to meet a growing demand; it also brought higher and higher budgets.

In the mid-sixties the main academic center began to perform an increasing amount of work for users at nearby colleges and universities. These users were charged the same rates levied against internal users. This arrangement offered useful services to the other institutions and provided incremental funds to support the center. In the early seventies we began to explore ways in which we could expand the concept of shared computer facilities.

The motivations for such sharing were primarily the desire to gain greater economies of scale in hardware, software, and operations. It was felt that a regional center, operated by professional managers with clear-cut revenue and cost goals, would provide cost-effective computing that could not be matched by each institution's own facility. By divorcing the center from the control of any single institution, we felt that each member would be encouraged to divert an increasing share of its computing dollars to a powerful computer utility. Each institution could

* The opinions expressed are those of the author and not necessarily those of the University of Pennsylvania.

then concentrate on the effective use of computers in education programs and administrative applications rather than getting caught up in the day-to-day problems of running a computer center.

Some Problems Encountered

In several ways the operation of UNI-COLL has met our expectations. The new organization has encouraged increased professionalism and attention to good business practice. The accounting for costs and revenues has become much more explicit than it was before.

We were not, of course, naïve enough to 'think that the new organization would solve all our problems of providing computer services. We were therefore not surprised to find some difficulties in moving toward shared computer facilities. It might be useful to describe some of these problems, since they are probably common to most such regional centers.

Cooperation among Independent Organizations

Perhaps the central difficulty stems from the very nature of the relationship between UNI-COLL and its institutional members. Each is a separate organization with its own goals. It is naturally the goal of the University of Pennsylvania and other users to get high-quality services at as low a price as possible or, alternatively, to get the best possible level of services at a given level of expenditure). UNI-COLL, on the other hand, must concern itself with its financial stability, the avoidance of unessential costs, and increasing its revenue. Some conflicts between supplier and user inevitably arise, at least in the short run.

Pricing

This problem is perhaps manifested most clearly in the pricing policies established by UNI-COLL management. The intent was to set prices so that the average job would be charged the same on the present IBM 370/165 as was charged on Penn's previous computer, a 360/75. Thus, none of the technological advance from the 360/75 to the 370/165—a reduction in cost per calculation by at least a factor of 2—was made available to users in the form of cost reductions (although benefits in the form of faster turnaround have certainly been achieved). In fact, some jobs are now charged more than before because their special characteristics do not allow them to take advantage of some of the new 370/165 features. UNI-COLL has not yet reached the break-even point, and so its management feels that it cannot afford to lower prices. Many users, on the other hand, feel that lower prices would encourage a sufficient growth in use to offset the reduction in rates. The issue obviously hinges on one's perception of the price elasticity for computer services.

Use of Idle Capacity

A problem that plagues all computer centers, but especially one that exists as an independent cost and revenue center, is the question of marginal pricing of unused capacity. The replacement of the 360/75 by the 370/165 resulted in an increase in capacity by a factor of almost 3. Demand has not grown nearly to this extent, and so we find ourselves with a machine on which more than half its available capacity goes unused. At the same time, we have a number of worthwhile applications that could use the idle capacity, but not at full average cost. If somehow we could use this capacity at something modestly in excess of marginal cost, everyone would be better off. So far we have not found a way to get around accounting conventions, government regulations, and UNI-COLL's reluctance to disturb its revenue base.

Commitment to UNI-COLL

Still another common problem to be resolved is the extent to which each participating institution should be committed to support UNI-COLL. This problem, of course, is faced by all university computer centers. To capture the economies of scale for all institutions, it is desirable that each member funnel most of its computing funds into a single large facility. On the other hand, individual schools or faculty members may find it more attractive to meet their particular needs through services other than UNI-COLL. The trick is to establish services and set prices so that individual users will be motivated to behave in a way that corresponds as closely as possible to the interests of the community as a whole. UNI-COLL, like most other computer centers, has not yet solved this problem.

Gaining Agreement among Many Organizations

Each participating institution, as well as UNI-COLL, has its own set of priorities. These tend to differ, depending on the balance between administrative versus academic computing, research versus teaching, and experienced versus inexperienced users. These differences ultimately come down to the question of which services should be offered, what price should be charged, and what priorities should be assigned to alternative uses of resources. Typical of these questions are the importance to assign to interactive versus batch processing, large jobs with setups versus short jobs without setups, and low prices versus extended services. These problems certainly exist within a single institution, but they tend to compound as one expands the variety of institutions served by a common facility. It is often difficult to obtain a consensus on an important issue, and so UNI-COLL may be forced to make a decision that satisfies no one completely.

The wide variety of services demanded by an expanded community of users places a special burden on the professional staff of the common facility. It has proved difficult to find managers with a strong background in administrative data processing, large-scale research computing, and unsophisticated student and faculty applications. As a consequence, it is difficult to serve each of these fairly distinct populations equally well.

Some problems have been experienced in deciding what services should be offered by UNI-COLL and what should be provided by the institutions themselves. For example, the operation of remote terminals, providing consulting services, or the technical evaluation of alternative terminal and communications configurations could either be centralized at UNI-COLL or decentralized to the individual institution. After a year's experience we have not yet fully resolved this issue.

Problems of Transition
Finally, transitional problems always arise when institutions attempt to change their mode of operation. The purpose of establishing UNI-COLL was to create a facility that could provide the primary source of computing services for most of the colleges and universities in the area. The shift to UNI-COLL has not happened to the extent hoped. Penn continues to provide well over half of UNI-COLL's revenue, and no institution has made a major reallocation of its computing funds from its own facility to UNI-COLL. There are nearly always technical problems in moving from one machine and operating system to another, even when the facilities are quite similar. Furthermore, the very persons who must be relied upon to implement the change often see the transition as affecting them unfavorably by threatening to reduce the control they now have over their own computer. They can almost always delay the move by arguing against it on technical or operational grounds. Even the simplest move, with the best will on the part of everyone concerned, is never painless; under less than ideal circumstances, an institution can always find grounds for delaying the conversion to a shared resource.

On hindsight, at least, it appears that the large increase in capacity that was installed in anticipation of fairly rapid growth is, for the time being, in excess of our combined requirements. Costs have thus been higher than justified by revenues, and so users have not been able to enjoy the full fruits of either technological advance or economies of scale.

Likely Long-Term Configuration of Regional Computing

The problems that we have experienced are not unique; they are probably common to any new regional center trying to evolve toward a shared facility that serves a wide variety of users. There is little doubt that the concept of regional computing is firmly established in the Delaware Valley. The task that faces us is to determine the best means of sharing common facilities and of providing responsive services to users. Based on our recent experience, a pattern of regional sharing begins to emerge.

Hierarchical Network

It appears to me that we will evolve toward a hierarchical computing network. At the bottom of the hierarchy will be specialized minicomputers and intelligent terminals serving each of the separate institutions. These local computers will provide specialized services such as interactive computing for the small user who does not need the powerful resources of the centralized center. They can also serve as communications concentrators for the local terminals, provide remote job entry capability for batch processing on the central computer, and handle on-line monitoring functions at the local level. The simplifications and economies of specialization offered by minicomputers often far outweigh any lost opportunities for economies of scale brought about by the diversion of resources from the regional center.

At the intermediate level in the network hierarchy will lie the regional facility. The primary emphasis will be placed on achieving great economy and effectiveness in handling batch-processing jobs. The regional center can also provide services to the relatively small population of users who require—and can pay for—the powerful hardware and software resources that can be made economically available only at a shared facility. Since many of the small, specialized jobs will be handled locally and will therefore never reach the regional center, the central computer need not divert much of its attention from its primary purpose of providing a powerful, generalized system. By combining the requirements for such a capability over the entire region, significant economies of scale can be realized.

At the top of the hierarchy will stand a national computing network.

It will be especially useful in providing specialized hardware and software services that provide substantial economies of scale. For example, one of the computers on the network might be specially designed to handle very large matrix manipulation problems. The fact that the computer can draw from the entire national demand for this service permits it to exploit economies of scale; the fact that it is tailored for matrix manipulation provides economies of specialization.

Ideally, the hierarchical nature of the computer network should be largely transparent to users. It should be simple for a user to transfer a job from one level in the hierarchy to another; this could even be done automatically in some cases. A program might thus be developed locally on an interactive terminal and later transferred to the regional center for routine batch processing.

Distribution of Operating Functions
The assignment of functions will follow a hierarchical structure similar to the hardware. Each institution will provide local services that fall within its field of expertise. The operation of the local terminals and minicomputers, the provision of consulting services at a relatively elementary level, and the determination of the local hardware and communications requirements will largely be administered by the individual institutions. In this way these services can remain responsive to local needs. Decentralization will also allow each institution to make decisions that recognize its own set of objectives and priorities without having to coordinate closely with all the other regional institutions. In general, the bias should be in favor of handling a function at the local level unless a clear advantage exists for handling it at the regional or national level.

The argument for regional sharing can be made in the case of a detailed service not specific to a given institution and not economical to support at each separate institution. For example, expertise in systems software is better provided at the regional center than at any of the individual institutions. Other examples of functions that should be provided primarily at the regional level are the evaluation and programming of generalized software, consulting assistance in hardware evaluation and selection, and consulting services for the smaller institutions that cannot justify maintenance of their own professional staffs. Ideally, the regional center should provide a common link to any external suppliers (including the national network), thus presenting local users with a single source for all services.

Price Structure

The pricing of computational services is extraordinarily complex. The intent of any rational pricing scheme is to allocate available resources in a way that leads to the greatest overall effectiveness and efficiency. To the extent possible, users' behavior should be motivated by the pricing mechanism rather than by artificial constraints (such as forcing them to spend their computational dollars on internal services).

A number of principles should govern the establishment of the price structure:

- Users should pay for the individual resources they employ (for example, the central processing unit, primary memory, auxiliary storage, input-output channels, specialized software packages, consulting services, and so forth).
- Rates for each resource should vary depending on the level of service the user requires and is willing to pay for. For example, a user wanting ten-minute turnaround should pay more than someone willing to accept overnight turnaround (or even totally interruptable service, with no guaranteed turnaround).
- Prices should be set to motivate users to impose a relatively even load on the system—for example, by charging less for off-peak use and by charging a disproportionately high rate for a large program that seriously affects other users during peak periods.
- Prices should reflect true economic costs, so that one class of user does not subsidize others. Thus, for example, a specialized software package should be charged on the basis of use rather than being added to general overhead.
- If the cost of providing a service exceeds the cost of a similar service offered by an external supplier (including a consideration of reliability and continuity), the service should not be offered internally.
- Prices should reflect any risk-taking transferred to users. For example, if a user guarantees to provide a given level of revenue to the regional center, he should receive lower prices (or additional services) for bearing the resulting financial risk.
- Users should be kept informed of prices and levels of service so that they can make intelligent trade-off decisions (for example, between price and turnaround time).
- Prices should remain relatively stable over time (perhaps being adjusted every six months to reflect changes in demand or technology).
- Prices should reward efficient use of a shared resource. For example, a user who supplies accurate predictions of resource requirements

(CPU and channel time, say) should be rewarded, since this allows the job-scheduling algorithm to achieve greater efficiency in a multiprogramming environment.

- Users with specialized needs (for example, elementary BASIC or very large compute-bound FORTRAN calculations) should not pay the full price of systems generality (for example, consulting services, an extensive program library, and so on).

- Each institution should be permitted to set the prices charged to its own internal users, irrespective of the rates paid to the supplier. Thus, the local institution may add a markup on services obtained in bulk from the regional center or national network in order to defray the cost of providing local support services. In some cases a markdown may be made in order to motivate users to take advantage of bulk capacity obtained under a revenue guarantee to the regional center.

No one has developed a pricing scheme that best meets all these criteria. It is clear, however, that any satisfactory scheme is bound to be complex. A great deal of research needs to be done on this topic, particularly in dealing with the special characteristics of network pricing.

Financial Commitment of Users

Closely related to the pricing issue is the question of the financial commitment each user makes to the regional center. Any contractual arrangement between user and regional center should recognize the following economic facts of life:

- Costs are almost completely fixed in the very short run but become variable after an appropriate lead time associated with a given form of capacity adjustment. For example, changing shift schedules of computer operators can be done weekly, while making a major adjustment in the hardware might take a year or more.

- Computational capacity is completely perishable. If it is not used when it is created, it cannot be stored for later use.

- Risk should be transferred to the party most able to control it or to bear it.

A form of revenue guarantee appears to me to be the most practical arrangement for meeting these criteria. An annual guarantee would probably be quite adequate, since capacity changes over a fairly wide range can usually be made within this lead time. The scheme might work as follows.

Each year each institution could contract in advance (recognizing necessary lead times) for a given level of services. The regional center would then set its capacity on the basis of the aggregate guaranteed demand. Modest surplus capacity should be made available to provide flexibility in meeting unanticipated demand and to handle variations in load over short time periods.

A user should be given two primary incentives to provide guaranteed revenue support. First, he should be granted a fairly substantial price discount for assuming the risk that he will not use all the capacity he contracts for. A discount in the order of 25 percent is probably sufficient to motivate users to set the guarantee at as high a level as they can reasonably expect to use. The second incentive should be a guarantee that the user will get the capacity he needs. The scheduling algorithm on the regional computer should be such that a user receives a high priority in obtaining his guaranteed level of capacity. Beyond that level no guarantee would be given; service would depend solely on whether the capacity could be provided without undue deterioration in service to other guaranteed users. Thus, each participating institution would have both a price and a service incentive to set an adequate level of guarantee. It should be noted that these are precisely the same incentives (and entail precisely the same risks) that each separate institution would have to face if it fixed the capacity of its own independent computer center.

The regional center should, of course, be allowed to sell surplus capacity after it has met its guaranteed commitments. The surplus could exist because of planned excess capacity or because an institution failed to use its guaranteed level of capacity. The revenue from the sale of such surplus capacity should be shared on the basis of each institution's guaranteed support and the amount of its guaranteed capacity it failed to use.

Payments to the regional center should depend in part on the quality of service provided. Payments should be reduced if, for example, turnaround fell below a given level for a given priority of service. Since the managers of the regional center are best able to control reliability and service, they are the ones who should bear most of the risk. Their incentive to provide quality service could come through a bonus scheme or from the freedom to spend some of the surplus revenue on projects of their own choosing.

Price and quality incentives should be the primary means of motivating use of the regional center. Each institution—and probably its various organizational subunits—should normally be allowed to spend their computational dollars anywhere they choose (recognizing, of course, any short-term obligation assumed through a contractual guarantee). After suitable time lags, then, the dollars would flow toward the most cost-effective sources of services. The major exceptions to this freedom should be when one user interacts strongly with other users. This would tend to apply most strongly to administrative data-processing applications that share a common data base or are closely linked with other subsystems.

Conclusions

There is little doubt that regional computing offers substantial advantages. The time is long gone when each college or university should try to remain self-sufficient in providing computing capacity. Quality of service and costs can both be improved through intelligent, well-managed sharing.

We face a very great challenge in implementing a hierarchical network that best meets our needs. Many technical and managerial problems must be solved before we achieve a completely satisfactory system. Success requires great skill, goodwill, and a keen sensitivity to the needs of the community as a whole. For a system as complex as a hierarchical network, success can be achieved only through an evolutionary process that adapts as changes take place in technology and our perceived needs.

CHARLES WARDEN
Data Resources, Inc.

21

An Economic Policy
for University
Computer Services

I am not here as a representative of Data Resources, Inc., nor do I intend
to talk about it.* I come, rather, as an economist to talk about the economics of resource sharing. Resource sharing is a thoroughly proper
theme for economics to comment on. Economizing, indeed, is the practice
of optimal use, not saving. We try to make that point by pointing out
that the man adrift in a lifeboat economizes his water by using it wisely,
not by saving it until he is rescued. The central subject of this essay, therefore, is how to economize on the use of computers—that is, how to use the
hell out of them, wisely.

In the process I shall say some nasty things about data-processing
managers. Therefore, let me say that I really do admire the kind of job
they have to do. It is among the worst assignments in the world. If they
do a perfect job—no crashes, no lost files, no delays in turnaround—they
are doing what is expected. Anything less is open to criticism.

Is the Computer Service Industry a Natural Monopoly?

George Feeney (Chapter 19) captured the spirit of those who support
centralized resource sharing by claiming that computer utilities will provide a superior alternative to fragmentation and inefficiency in today's
decentralized computing, even as electrical utilities replaced established
decentralized power generating seventy years ago. It is important to
examine this parallel and, I believe, to reject it.

Electricity is a homogeneous product. Capital costs for generation and

* Data Resources, Inc., is an unfunded, profit-making company specializing in economic forecasting, model building, and business consulting through time sharing. It
provides economic data bases, encompassing over 200,000 data series, statistical programs, forecasting, models, model simulating programs, forecasts, and the supporting
technical services to go with them. Founded in late 1968, it is today the largest commercial Burroughs installation in the world, with four B5500's and one B6700 with
400K of memory. The time-sharing system is supported by a dual telecommunications
network: one, maintained by DRI, is based on multiplexers and covers seven major
cities: New York, Washington, D.C., Pittsburgh, Chicago, Los Angeles, San Francisco,
and Seattle; the other is Tymnet, Tymshare Corporation's digital system, which covers
approximately fifty cities. Although nationwide by geography, DRI is "regional" by
application and serves a limited community of users, composed largely of economists,
macromodel forecasters, management science and operations research people, market
researchers, business planners, financial managers, stock analysts, and industry researchers.

distribution dominate and, importantly, there are ever-increasing economies of scale almost as far as the eye can see. There are large externalities: social costs from duplicate suppliers such as would occur when four or five electrical companies put up separate electrical poles in the community.

The computer services industry does not have these utility features. The minimum optimal size of a computer facility is quite small, and there can be a lot of them. Industry cost ratios do not seem to vary much with level of output. It is true that an individual facility frequently can distribute more of its services without incurring more costs. It can expand its output locally with only marginal increments in the cost. But the key point is that large facilities overall do not display greater efficiencies than medium-sized facilities. The distribution of costs for such categories as hardware, telecommunications, sales, and services is remarkably stable across a wide range of facility sizes. It is also clear that the profitability of facilities is not closely related to size.

Demand curves for computer services are not inelastic in the short run and elastic in the long run, despite popular opinion. Quite the contrary: they are very elastic in the short run, because the activity of conventional users rises sharply if the price is lowered. Even in a restricted market like a university computer center, lowering the price initiates greater use. In the long run it is not quite so elastic because the decision to use the computer services is related more to nonprice considerations like the expectation of new services and new applications than to price. A company will decide to increase its level of computer services in five years if it has expectations for a whole new level of activity to computerize, but it will not do so simply because of a 5 or 10 percent change in the price.

Finally, unlike the case of a public utility, computer services do not embody an overriding public interest that must be protected. Mistakes will happen; programs will have bugs, and computers are going to crash. But it is not clear that we should encourage regulation to protect the public interest.

Computer services therefore have important differences from the usual image of a utility. Because of these differences I think it is mistaken to argue that computer services will follow naturally in the path of concentration that made electric generation and distribution a public utility. If it were true, I do not think it would be good. Feeney seems more concerned about the problems of brownouts and telecommunications than about algorithms, cash flow models, regression techniques, multiple use of files, and other problems of providing better computer services. He is worried about the tool parts of the services that are provided by utilities, not the things that are done by all these free-wheeling programmers working at ten or twenty dollars an hour.

But even though the computer services industry is not a natural monopoly and is not a likely candidate for making a public utility, unwise public or quasi-public policy could force some monopoly traits on it. Therefore, we should examine what implications that development might have.

Should the Computer Service Industry Be Made a Public Utility?

The economics of resource sharing can be viewed as a question of industrial organization. It ought to be looked at as a service with a marketplace. We should be concerned about such things as market structure (what is the concentration ratio?), behavior of firms (will there be an effort to minimize costs, or will there be some conspiratorial, oligopolistic arrangement?), and industry performance (will the industry give us what we want?). We should look at pricing to see how consumer preferences are translated into votes. In the open marketplace the ballots are dollars, but that is not the only way to organize the market. There ought to be some sort of market discipline. If declining marginal utility and increasing marginal costs do not provide the combinations for effective market discipline, there must be some peculiarity in this marketplace that requires us to think of it as a utility—some technological relationship, for example. Then the market structure question becomes critical. We have to think of the industry as a public utility, and we have to establish a public authority to exercise control for the public's interest. Stating this conference's central question as how to set up a network to share computer resources builds in the inference that organizing, structuring, establishing common protocols, vesting control and authority, and otherwise centralizing is the way to serve the best interests of the university community seeking computer services. I do not think that is the only or the best answer. The issue is not how to share computers or communications. The issue is, I think, how to guarantee the best buy of computer services, where best buy is a combination of many different measures such as quality, quantity, price, variety of choice, continued technological progress, equitable distribution, efficient use of resources, stability, reliability, and freedom to compute or not to compute. These are the criteria for a proper economic policy toward computer services or any other industry.

What does our experience in other industries tell us about the structure, behavior, and performance of an industry? It suggests that the consumers' best interest on these best-buy criteria is served poorly by each step that reduces competition in the marketplace, though the interests of the surviving suppliers are indeed richly favored. Not surprisingly, therefore, suppliers are motivated to reduce competition, often with proposals that on the surface are seemingly congenial to the public interest—such things as certification procedures to ensure quality, licensing to eliminate quacks, and fair-trade pricing and regulated rates to protect against unfair price gouging. Proposals to overcome these evils are usually devices to establish barriers to entry to protect the current suppliers from the challenge of new suppliers with a better mousetrap.

Unbridled competition may not always be best. Exceptions to the rule that competition is best are found, or at least often cited, in those industries where some condition exists to make the controlled monopoly appear preferable. Two conditions are often proposed in a political economy course to qualify an industry as an exception: One is where economies of scale are present so that the average cost per unit declines over the full range of potential output, suggesting that efficiency is best achieved through a single supplier. Electric utilities and communications are familiar examples. The second is where the consumer is both highly vulnerable and terribly inexpert. Health care is usually cited as an example of the second condition.

Tom Kurtz observes in Chapter 30 that there is no role for a broker of computer services such as we see in real estate, insurance, or the stock market. This suggests to me that the consumer is quite expert and does not need special help in developing information about the market for computer services. Evidently he can bargain directly and effectively with the supplier. He can test and choose among different report writers, statistical packages, even different programming languages quite readily. Therefore, the second condition is not met.

I conclude, therefore, that the computer services industry is neither a natural monopoly nor a special industry that requires regulation for some overriding noneconomic reason.

Competition in the Computer Services Industry

Indeed, it would seem that the computer services industry, up to now, has behaved in a remarkably competitive fashion. Further, its performance suggests that it is in excellent health and that any change in structure leading to a decline in competition would be detrimental to the consumer of its services. Let me cite some of the characteristics that I and others, as suppliers of computer services, abhor and would like to eliminate. But that is because I am a supplier and do not like competition. Suppliers like a protected market, where they can operate without challenge. From a social policy point of view, however, it seems to me these are excellent virtues. Let me run down a list of some.

1. *Great freedom of entry.* Neither technical nor financial barriers prove much hindrance to new groups to provide computer services.

2. *Ingenuity in solving marketing and distribution problems.* The ingenuity is most evident in dealing with low-quality and deteriorating communications services. And, by creative marketing, these newcomers tailor their services to find a place in the marketplace.

3. *Innovative behavior.* A great flood of both hardware and software development continues, as advertising in the trade journals and at conferences richly testifies.

4. *Abundance of choice.* There is no shortage of potential suppliers. At one count there were fifty-five vendors competing in New York City for service bureau business.

5. *Transitory monopoly positions.* Although innovations can provide a temporary monopoly, there is a minimal ability to sustain it. The absence of any copyright or patent protection further weakens this market control.

6. *Low profit margins.* Thin profit margins and, frequently, the absence of any current profit at all lead to a keen cost consciousness.

7. *Variable supply.* There is rapid expansion of supply in periods of expanding demand and periodic purging of inefficient suppliers during bad times, as in 1969 and 1970.

8. *Resource mobility.* Not only are the computers readily shifted from one application to another, but the labor force floats very easily from one employer to another with none of the language, emotional, financial, or technical handcuffs of some industries.

9. *Low concentration ratio.* Nobody has over 5 percent of the market.

10. *Wide spectrum of suppliers.* There is participation by suppliers whose primary activities are in widely different industries. The telephone companies are in it, as are banks, insurance companies, manufacturing firms, and aerospace companies. Clearly, there is an ability to pop in or out of the market as opportunities change.

11. *Socially desirable pricing.* There is a relative absence of discriminatory pricing and internal subsidies, which is excellent from a public policy point of view and certainly an important thing to keep in mind.

12. *Ineffective vertical integration.* None of the hardware vendors has been able to capitalize on his manufacturing position to dominate the market. The ruling against IBM and SBC may have helped this a lot. But unlike the classic cases in the aluminum and petroleum industries, making the basic hardware does not seem to give computer manufacturers a substantial competitive advantage in the computer services industry.

13. *Lack of dependence on suppliers.* There is a low level of control from suppliers, with IBM and Bell Telephone being the two big and serious exceptions.

14. *Lack of monopsonist control.* The suppliers are not dealing with a single client, in most cases, and certainly the market is not dominated by a single major consumer as, say, the defense industries are or as the automobile parts producers tend to be.

All in all, these are the characteristics that lead to a healthy, vital, and energetic market. It is a list of virtues to be jealously guarded by the consumer, for it is a list of evils to suppliers like me, and we will attack them at every chance.

These characteristics, which have made the computer services industry highly competitive and often chaotic, are the very characteristics that have given rise to the fantastic progress, high quality, and wide variety that we have enjoyed in the last decade.

The University Computer Services Market

The obvious question is, why cannot a university enjoy the fruits of this market, too?

But it is better to ask, how many of the above virtues exist in the university computer services market? Universities have succeeded where business has failed in isolating the market, shielding it from entry, establishing vertical integration, being captured by suppliers, and replacing quality control through consumer choice with medicine men and institutionalized ritual for allocating computer time. As a final crowning achievement, they have succeeded in banishing the ruling coin of the realm (dollars) and enthroning in its place monopoly funny money.

The results are not surprising. Costs have gone up, quality has fallen behind, adoption of new techniques has been slow, and the broad base of users has been miserably serviced. True, the high priests have fared well with high salaries, positions of honor, and a deepening aura of mystery about their incantations. But the consumers of the services have agonized with precious little relief. Indeed, so bad has the situation grown in some universities that decapitation was preferred to lobotomy and the whole computer center was disbanded.

This is, of course, a scenario in absurdity, for the universities should be, and have been in the past, the source of innovation and discovery. But today, without an effective market pressure, that creative spirit is not active.

It is to this end, rejuvenating that creative and imaginative environment in university computing services, that the National Science Foundation should bend its efforts. The NSF should encourage the "best buy" in these services. It should stimulate choice, not restrict access. It should encourage a multiplicity of suppliers, not institutionalize a monopoly.

If the NSF is to support the university use of computer services (and it is by no means clear that it should), the support should go to the user, not the supplier. Let the consumer shop for his services freely, so he can keep the feet to the fire at the computer facility. If his own university center cannot do the job, let him go outside, to other universities or to the commercial market. By all means untie his funds and forbid universities from forcing him to use the services of an in-house computer. And, importantly, beware of the internal subsidy that reduces the cost of computing by transferring the proper share of overhead and other costs to others in the university community.

Indeed, if the NSF hopes at all to improve the quality and sophistication of the software and services available to the research community, it must abandon the support of software development groups. Such support only installs dictators unresponsive to user requests. Give the support to the consumer so he can get his needs listened to.

On the surface this may seem a poorly cloaked pitch to get a piece of the university market. Let me uncloak it by saying I want to service that market and think Data Resources has a good service for it. But there is no way I should be able to compete effectively with the tax-free, low-wage, nonprofit, and debt-free university suppliers. If I can, then they deserve to be put out of business.

Implications for a University Communications Network
The proposals to subsidize a university communications network is one more wrinkle in the cloak around the university market. Communications cost will never be a major part of computing cost. Indeed, most suppliers of computer services are willing to absorb those costs and find ways to minimize them through concentrators and the like.

And, as in the computer services market, there are a number of com-

munications suppliers, and new ones are appearing. I refer you to
Packet Communications, Inc., 214 Application, P.C. 8533 as an illustration. The emergence of a number of regional networks is more evidence
that such efforts can be made cost effective without direct federal subsidy.

However, once again, if the NSF seeks to help users overcome communications barriers to the broader market of computer services, it
should offer its support to the user, not the supplier. Where there is effective demand in the hands of the user, suppliers of these services will
arise en masse.

22

ZOHRAB A. KAPRIELIAN
University of Southern California

The Politics
of Cooperation

I want at the outset to give some background on the resource-sharing project in which I have been involved at the University of Southern California. This description will provide some understanding of the experience upon which I will base some subsequent comments on the topic of this paper, the organizational and political considerations of computer sharing.

In late 1971, the chief executive officers of three major institutions in the Los Angeles area—the California Institute of Technology, the University of California at Los Angeles, and the University of Southern California—met and decided that it was appropriate and timely for their institutions to examine the potential mutual benefits available through sharing of resources. A vice-president was designated at each university to lead these cooperative studies. Three areas were identified for further detailed evaluation: computing, library services, and academic programs and facilities.

Computing was selected for first attention. A target date of approximately six months was set for reporting the initial results. By February 1972, with National Science Foundation support, a major feasibility study was under way. Concurrently, in-depth studies are proceeding in the library and academic areas. We have additional financial support from the Carnegie Corporation and the Rockefeller Foundation. Our progress to date indicates that these efforts will produce significant and worthwhile results.

The study in the computing area sought to analyze the feasibility of coordinating and pooling any or all of the computing facilities and services at the three universities. To assist us with this study, we engaged Arthur D. Little, Inc., as consultants to work with us for approximately six months. A thorough effort was made to search out any serious impediments to proceeding with the concept of sharing. A number of different aspects of feasibility were carefully examined, including organization, procedures, technology, finances, and management. The consultants' final report concludes that the pooling of computing resources among these three universities is feasible on all counts.

The report is currently being reviewed on our three campuses. In very general terms, we have concluded that the most effective arrangement for computer sharing will be a distributed network making the computing facilities and services on each campus available to the others through a communication link. Our concerns now are more financial and organizational than technical. This is not to minimize the importance of the necessary technical work. We think there are some very interesting and challenging technical questions, including the degree to which the network can be made transparent or invisible to the users, but we are optimistic that these questions can be solved. Our main concern (because there is a lack of experience or suitable models) is the broad area of management, including fiscal and organizational considerations.

What we can learn from experience elsewhere, such as the Advanced Research Projects Agency network and installations such as Triangle Universities and the Merit Computer Network, makes us confident about the technical feasibility. Experience elsewhere also indicates, however, that we need a high priority on early solution of the nontechnical questions. We are concluding that the organizational and financial questions should and can be resolved first, or at least concurrently with the technical implementation. It is apparent to us that we cannot hope to build significant volume of work load on the network until we have answers in the financial and management areas that are acceptable to the participating institutions.

To help focus on the organizational and political implications, we might divide these into internal and external considerations. The internal questions are fairly obvious; for example, of our many constituents, who should be involved, at what point in time, for what purpose, to what degree, and in what depth? Any university interested in computer sharing will need to decide the appropriate involvement of, for example, board of trustee members, administrators, key faculty members and committees, students, computing center directors and their staffs, possibly external consultants, and perhaps others.

Our experience to date indicates the value of early involvement and counsel of senior faculty members who represent the highly sophisticated computer users.

Another important internal consideration is the role of the computing center director and the degree to which he represents a vested interest. This is further complicated if you are involved in sharing in other areas, raising questions as to the relationship of these efforts to computer sharing. For example, what is the proper internal relationship of the computing center organization and management with those of the consortia, and the relationship of both of these with the central administration in general? Just to illustrate the complexities, we now are reviewing half a dozen alternative organizational structures that might be appropriate.

Regarding external considerations, a chief question is: who should represent the participating institutions in a network? Who will determine basic policies regarding sharing risks and costs, network management, conflict resolution, membership policies and commitments, and relationships with other networks? These are key considerations and warrant very early and careful attention.

I believe the organizational and political questions, both internal and external, are fairly well known. Knowing and understanding the problem is, I hope, the first step toward solution. Let me move, then, to what I feel might be some basic ingredients for successful sharing.

In simple terms, we might start with the assumption that no one prefers "sharing." Sharing usually means giving up something. Webster's definition states: "Share implies that one as original owner or holder grants to another the partial use, enjoyment or possession of a thing. . . ."

The fundamental question of whether we are really interested in sharing probably revolves around the availability of resources. If we can have our own thing because there are plenty of them or because we have sufficient resources to make or buy our own, then we are probably not seriously interested in sharing with others. It appears that we turn to a consideration of sharing only when we understand that there are not sufficient resources to have our own thing.

Although I recognize that there is interest in computer sharing because networking is new, exciting, and on the forefront, I also feel that serious consideration of broad-based sharing in higher education will be expedited when it is more clearly understood that we cannot meet our goals by continuing on a completely separate and independent basis. Our natural reluctance to share will be overcome when and if sharing is seen as the most viable alternative by which we can attain our objectives.

In light of these generalizations, let me summarize several points that might be considered as ingredients for successful sharing:

1. The need to share must be understood. I think the financial crunch in higher education is helping to make this more apparent.

2. There must be mutual benefits. Generally speaking, we will give up something only to gain something. At our three institutions, we are starting in computer sharing from positions of strength at each campus. There is something here for each of the others.

3. There must be good faith and trust by the participants. It should be

understood that we probably cannot reduce everything to contract form or document all details.

4. There must be intelligent management. Worthwhile sharing in higher education will be a real challenge, but proper management can solve the technical, financial, and organizational issues involved.

Let us assume the ideal situation and the presence of the essential ingredients: the need, mutual benefits, good faith, and intelligent management. What are the political challenges with which this intelligent management must cope successfully? Let me illustrate a few:

1. *The politics of financial responsibility and accountability.* If we divert part of the funds now supporting on-site computing to network computing, we can easily generate opposition from faculty, researchers, and students. We all know the typical questions:

- Are we undermining our own facilities to support someone else's?
- Are we giving up our scarce resources and dollars to support something that we do not fully control?
- Is someone else benefiting at our expense?
- Are there unknown fiscal traps or unrecognized financial risks?
- Would not we be better off on our own?
- Are we sure we are getting the most for our dollars?

2. *The politics of highly technical considerations.*

- How can the decision makers understand the technical implications?
- Who will assure network capacity, reliability, stability, and so on?

3. *The politics of institutional goals, and state and federal implications.*

- How does computer sharing fit with our academic objectives and goals?
- How compatible is general computer sharing with the desire for institutional excellence?
- What are the implications of private versus public institutions, or statewide public networks versus regional heterogeneous nets? Or of state and federal political mechanisms?

From these few illustrations, we might conclude that although many of the political implications of resource sharing are fairly obvious, they can be complex and their proper solution can be difficult.

In spite of the difficulties of the task, we are optimistic about our cooperative efforts with Caltech and UCLA. Some of the aspects I feel will help us include the following:

- Our project has top-level support and is under the direction of three senior vice-presidents.
- The three universities are essentially equal participants, each sincerely interested in potential mutual benefits.

- We are attempting to do a thorough feasibility study, with careful evaluation of alternatives before proceeding with implementation.
- We are examining several basic and interrelated areas: computing, library services, and academic programs. This gives us opportunity for some trade-offs; for example, one university with greater strength in computing may be relatively weaker in library resources.
- Our emphasis is not on cost reduction but on longer-term improvements, including broader and better services to our faculty and students, and improving the return on invested resources.
- We have solid experience in computing at all three universities; by any standard we are large and sophisticated users. Our current expenditures are approximately $6 million per year in the three centers we are considering linking together, and I should point out that this is only part of the present computing at each campus.
- We have good experience to date in the technology of computer sharing, including active participation in the ARPANET.
- Our key faculty and sophisticated users are not only interested but are excited about the possibilities that might be opened up through networks. Their cooperation and support will be most helpful.

In conclusion, we are encouraged by our cooperative effort to date. We recognize that there are risks in sharing, but the potential mutual benefits appear to outweigh the possible disadvantages. The political implications are there, and must be recognized and dealt with successfully. I am confident that this can be done.

23

CHADWICK J. HABERSTROH
University of Wisconsin—
Milwaukee

Behavioral Implications of Organization Change

The adoption of a large-scale national computer network will pose many organizational problems for member universities and other research organizations. The network itself will be an external institution over which the member organizations will have little if any control. The problem thus becomes one of adapting to an external force and and finding ways of exploiting the opportunities it makes available. At the same time the organization must maintain its internal integrity and and continue to cope with its principal mission and the remainder of the relevant environment.

The aspects of the network environment that will pose the central organizational problems are (1) the changing technology in the field of automatic computation, which appears to result in a four-year cycle of technical obsolescence, and (2) the externalities that arise in computing because of the institutional growth of networks.

With new and superior hardware and software systems being developed regularly, no user can afford to ignore for very long the renewal of its physical capital, the retraining of its personnel, and the rethinking of its information-processing applications. Each of these represents a capital investment that must be amortized over a period of four to ten years. The network itself can absorb a substantial part of this reinvestment activity on behalf of the constituent organizations, thus reducing the pressures on the individual institution and spreading considerable parts of this overhead over the entire set of participating institutions. This is one major source of the externalities that can be exploited in a network. The other important source of externalities arises from the fact that the optimal size installation for some applications is big enough to service the entire nation and thus is well beyond the scope of any individual user organization.

The problems in exploiting these externalities arise from the need of member organizations to adjust their own computing systems to link with the network. This imposes special investment demands that are not recoverable except in the context of successful use of the network. A relation of dependency is established, with the effectiveness of a substantial portion of the member organization's computing budget being determined by network activity that is not under its own authority or control.

Effects on Organizational System Integrity

Any organization develops a pattern of operation that is familiar to the individual members and forms the basis for their cooperation. The key aspects of this pattern are the status system, the major functions, the subsystem reifications, and the specific work roles.

The functional structure differs from organization to organization. In a university it usually consists of financial management, admission and records, student services, and the academic schools and departments. In universities the computer center has typically been set up as an independent function, whereas in industry it is more likely to be part of the financial or accounting function. The reifications that are of importance to the problem of networking are primarily the computer center, the network as an external institution, the other functional departments, and the university or other organization as a whole. The specific work roles that are critical are those of technologist, machine proprietor, operator, programmer, systems analyst, consultant, and user. These roles seem to be common among all the relevant organizations.

It must be emphasized that the specific pattern of the member organization represents the deepest level of commitment. It is not merely a matter of linguistic convenience or an exercise in the drawing of organization charts. It is the central determinant and constraint on the status and rewards of the individual participants and on their job security. It is also the embodiment of the organization's central mission and of its particular competence. This pattern exists to serve central purposes of the organization and its members. In any ongoing organization it has survived a wide variety of past environmental crises and will be vigorously maintained in the face of future ones, including those posed by the development of external networks.

The network link is primarily relevant to two of the major reifications: the computer center itself at the technical level, and the top organizational management at the strategic or entrepreneurial level. The pressure inside the organization that will work in the direction of network involvement will arise from the roles of the technologist and the user. The technologist will be attracted to networking simply as a technical challenge, will wish to use and exploit the network merely because it is there. The user, on the other hand, will see opportunities of accomplishment in his field that cannot be realized without the involvement of network resources; furthermore, the use of more sophisticated machinery and software will reduce development effort on his part and will stretch his budget dollars. Unfortunately, these are the only work roles that can be counted on for support of network development. To all others, networking will mean new demands, disruption of their established work patterns with no accompanying intrinsic rewards. It is from these other roles, acting in the name of the organization's mission and traditions, that resistance to networking can be expected.

Many conflicts in organizational purposes will have to be worked out in the process of exploiting a national network. The network will necessarily link many unlike organizations. Some in the business sector will be operating under profit constraints. All will have goals of research output. Many will be educational institutions; others may be providing services or products as well.

Within each organization, the organizational pattern reifies many specialized capacities that are needed to achieve its own purposes. A more or less stable status order among the specialized capacities and people will be exposed to adjustment under any adaptation as basic as will be required for full exploitation of computer networking. The relative position and cooperation of specialized competences are controlled by established measures of performance that define mutual expectations in reasonably precise quantitative terms. Budgets and variances are among the most important of these tools; budgeting for network activity will necessarily reduce the resources available to other functions. This change will produce conflicts, since the gains from a changing budget rarely accrue to the subunits that are required to make the sacrifices.

Another factor is that technical measures of performance such as machine downtime, percentage of utilization, and turnaround time typically govern the day-to-day functioning of a computer center. But there is no direct relation between these indices and financial controls. A radical change in the mission of the computer center would disrupt all the existing arrangements by which its staff members coordinate their activities. Empirical research on organizations has frequently demonstrated that such performance measures often work at cross-purposes with organizational missions, even after years of stable operation and abundant testing of both the mission and the technology. It will be much more difficult, then, to assess how the present procedures of an organization's computer center will facilitate or interfere with the demands of networking. In this area critical contributions in technical management will be needed.

So far I have said little about the human problems of organizational adaptation. The organizational forces just discussed express themselves mostly through the decisions of individual members who happen to occupy the key positions. It has often been demonstrated that the system of rewards dominates the decisions of individuals in the work context. Most reward decisions in organizations have already been made, and a very substantial proportion of the resources of the organization is completely bound into this reward system. All those resources that represent agreed material remuneration for individual participants remain untouchable. Everyone in the organization shares a need for a personal security in these rewards, and any interference with legitimate expectations raises problems that far outweigh any gains from budgetary flexibility. This explains the emphasis on reduction of work force through normal attrition rather than dismissal, even when dismissal is legally possible.

Less frequently recognized is the need of working personnel for stable personal relationships with the other significant actors in their work lives. In part this feature reflects the human need for a stable, well-structured environment, the human environment being the most important dimension. In part it reflects the long-term growth of relationships of friendship and cooperation, which are referred to as the informal organization and which represent an important matrix of organizational functioning. People also become personally invested in their positions of relative status in the organization and in the opportunities provided for personal achievement. They will eagerly accept job changes that open new avenues of achievement or enhance their status and will resist those that destroy opportunity or threaten status. Opportunities for personal growth are also held to be desirable in organizational life, even though only some individuals at some times may be able to take advantage of them.[1]

The rewards of individuals and their ability to influence the outcomes of organizational activity depend heavily on their place in the pattern of activity. This structure of expectations has two interrelated dimensions. One is the nature of the work, what is referred to in the organizational literature as the task model. It includes the goals of the organization and the programs that have been elaborated to achieve them. The other side identifies the key individuals and their relationship to the task model. This provides a *Who's Who* of the organization and is usually conveniently expressed in the standard organization chart or in greater detail in the organization's telephone directory.

The power of each incumbent depends heavily on his position in this pattern. The sources of power, according to the most widely accepted analysis, are (1) physical force and security therefrom (expressed in the computer context as control over equipment), (2) reward power (discretionary funds, not bound funds), (3) expert power (knowledge and skill), (4) professional and personal links in the human system (referent power), and (5) organizational legitimacy (the acknowledged right to choose from among alternative courses of action).[2]

From the preceding analysis we can discern several dimensions of organizational technique that would enhance the transition to networking. Vesting the control of equipment in technologists and users would facilitate this effort, as would giving them discretionary funds. The mobilization of expert power means the employment or retraining of personnel in network-related knowledge and skills. The use of the remaining two categories of power is less obvious. The informal organization is not subject to direction and control. It is responsive to problems, both those of exploiting the network and those of resisting it. Persons with legitimate managerial power are able to back networking or to fight it. Their responsibilities not only extend to the exploitation of the opportunities available but also include the maintenance of the organization and the achievement of other goals. The most that can be expected from key members of top management is a mild blessing from time to time and facilitation of the other arrangements. The key problem from the standpoint of the sponsors of networking is to maintain the alignment of the necessary competence, the reward picture, and the social relationships. Beyond that, if the claims of networking are valid, the movement toward this system will carry itself.

Perceptions of Adaptive Change
As indicated earlier, an adaptive change is a problem-solving activity responsive to change in the environment. Organizationally the key question is, "Whose problem?" The direct impact of the environmental change usually occurs at the entrepreneurial level. A frequent source of stimulation is the observance of a changed pattern of activity on the part of similar, potentially competing organizations. This situation immediately gets top management attention.

Another entry point is in the activities of the entrepreneurial researcher within the organization who represents the potential user of network services. He may invent an opportunity to exploit newly available resources, or he may imitate. The remaining category is the computer technologist who sees the opportunity to expand his competence and position. The entrepreneurial role undertaking the exploitation of network opportunities will be viewed as the sponsor of change within the organization. His friends and dependents will take this as a facilitating move, his habitual antagonists will probably view it as a threat.

Change activity impinges upon numerous other problems as it progresses. The most important are conflicts with established purposes. If resources must be allocated to the exploitation of the networking opportunity, other programs must make do with less. If change occurs in the context of organizational growth, these problems are much reduced. The reallocation of budgetary resources merely represents the shifting of potential to the new area. But change superimposed on stability or retrenchment will demand sacrifices. These may well represent the freeing of bound resources, which probably ranks as the most stressful act an organization can undertake.

Change always influences the vested interests of an organization. At the very least, the structure of expectations must change. All participants in the organization must relearn some aspect of their involvement. Likewise the reward structure will change to the betterment of some and the relative detriment of others. The needed balance of skills and knowledge will change, perhaps resulting in obsolescence for some previously valued capacities. Where a threat to vested interest is perceived, that interest will be mobilized as resistance to change.

The design of implementation activity, especially in the early stages, is a critical matter. A poorly conceived or halfhearted effort will generate problems much more rapidly than solutions. The first consideration is the time boundary: when change should begin and by when it must be completed. Premature action creates problems that will require a long time to work out; excessive delay can result in a lost opportunity. Few changes require involvement of the entire organization; yet if the change is conceived too narrowly, unanticipated consequences will affect those departments that have not participated. These can be costly. Too broad an involvement incurs costs in attention and worry that may be totally unfounded.

Those who undertake the implemention of change must commit themselves to see the matter through. This obligation implies responsibility for and effort toward the resolution of all conflicts that occur as consequences of change. It requires assistance to decentralized subsystems of the organization in adjusting to the changes that are imposed on them from outside. Adjustment opportunity must also be provided to co-operating independent systems, such as important suppliers and clients, as well as affected employees.

Manifestations of Resistance to Change

The key arena for effective implementation of change and for its resistance is the informal organization. Formal authority is generally obliged merely to recognize and consolidate that which is actually determined operationally in the working out of felt problems. The most immediate manifestations are a withholding or distortion of information that may be needed in the change and a failure of cooperation when and where needed by the sponsors of change. If problems are perceived as serious or directly threatening, or if the sponsors of change are perceived as intractable and unwilling to help in the inevitable adjustments, resistance may take more active forms of counteradaptation of direct sabotage.

Dealing with Resistance

In the environment in which all networking will be carried out (that is, a dynamic, rapidly changing technology), an organization cannot regard adaptation as a matter of crisis. It will be a recurring phenomenon. This fact imposes a need for and a great advantage to adequately planned change. Considerable knowledge is available to assist in this activity.[3]

Changing technology is not unique to the computer industry. Radically shifting markets have been known in many economic sectors for a long time. In recent years the radical shift of social values and opinions from liberal to reactionary to radical and back to conservative has kept concerned managers of organizations busy with adaptation. Many classic organizational mistakes can be recognized in the process of coping with these changes. There is no need to repeat them in the context of networking:

1. Any change activity represents an investment. It requires resources beyond those available for normal operation, and these must be amortized over some future period of stable operations, which is relatively short in the computer area.

2. There is no reason to assume that the specialized competences already available to the organization will be adequate under the changed conditions. Change may require the importation into top status positions of new talents.

3. Change requires many key decision makers to give up business as usual and to rethink their activities in the context of the entire organization's functioning. This effect may be referred to as the induction of global rationality (as contrasted to local rationality within one's own function).

4. The broadest level of participation in the change must occur at the earliest stages. This involves the entire informal organization in diagnosing and responding to the change and the needs that underlie it.

5. There must be an expressed and demonstrated willingness of the sponsors of change to adjust their design and implementation process to the human and organizational needs that become manifest.

6. The successful management of change requires a de-emphasis of status differences within the organization. Cooperation will be required between status levels that have not previously needed to communicate. The de-emphasis itself will make the acceptance of status changes somewhat easier.

7. Individual members will be more highly threatened if they find themselves trapped in some inescapable and uncongenial situation. An organizational response is to establish patterns of easy mobility among work groups and jobs and into and out of the organization. Such activity will place heavy demands on the personnel function in finding adequate positions for disaffected employees elsewhere, in recruiting the new talents needed, and in arranging intraorganizational transfers that will be generally satisfactory. Generous severance pay can help greatly in inducing voluntary departure.

8. People involved in a significant event need a feeling of common fate and a similarity of involvement. Emphasizing the ultimate favorable consequences of change for all involved and allowing all to share in the general design of the change process can help.

9. The change process must not impose unbearable strains. In particular, the process itself should yield satisfying social interactions to the participants.

10. Auxiliary help in the form of professional change agents can be employed. These agents provide the increased communication capacity that is urgently needed in the implementation of major change. They also represent a specialized competence that can facilitate problem diagnosis at the human level. As consultants they can serve as a source of ideas for coping with the human aspects of implementation problems.

Many of the preceding features of organizational design that can facilitate recurring cycles of adaptation have been explored in the organizational literature. The key reference is the contrast between mechanistic and organic forms of organization.[4] The mechanistic form represents conventional bureaucracy, which is considered well adapted to stable market and technological conditions. The organic form, which was first studied intensively in the electronics industry, seems better adapted to conditions of rapid environmental change. The relevance of the organic form to the area of automatic computation is fairly obvious. Table 23.1 is an abridged version of the contrast between mechanistic and organic forms.[5] It summarizes many of the points that are central to the preceding analysis.

Table 23.1. Contrasting Characteristics of Mechanistic and Organic Forms*

Mechanistic	Organic
1. Functional breakdown of whole task into specialty segments.	1. Contribution of specialties to whole task.
2. Commitment to Function (means).	2. Commitment to whole task.
3. Integration via hierarchy.	3. Integration by continual readjustment through interactions with others.
4. Precise definition of authority.	4. No limitation on authority or responsibility.
5. Limited responsibility commensurate with authority.	5. Broad commitment to organization and to tasks.
6. Hierarchic structure of control, authority, and communication.	6. Network structure.
7. Full knowledge of the situation only at top, reinforcing hierarchic control.	7. Knowledge diffused, any location may become the ad hoc center of control.
8. Vertical interaction.	8. Lateral and diagonal interaction.
9. Operations governed by instructions and decisions of superiors.	9. Communication consists of information and advice.
10. Loyalty and obedience are primary conditions of membership.	10. Commitment to tasks and "technological ethic" more highly valued.
11. Localist orientation.	11. Cosmopolitan orientation.

* Reprinted with permission from *IEEE Transactions on Engineering Management*, IEEE, New York, EM-15, March 1968, page 20.

References

1. See especially the writings of Abraham Maslow, whose theory of a hierarchy of needs forms the basis for the preceding analysis. Abraham H. Maslow, *Motivation and Personality*, Harper, New York, 1954.

2. John R. P. French, Jr., and Bertram Raven, "The Bases of Social Power," in Dar-

win Cartwright and Alvin Zander, eds., *Group Dynamics,* 2nd ed., Row Peterson, Evanston, Ill., 1960.

3. Warren G. Bennis, Kenneth D. Benne, and Robert Chin, *The Planning of Change,* Holt, Rinehart and Winston, New York, 1969.

4. Tom Burns, and G. M. Stalker, *The Management of Innovation,* Tavistock, London, 1961.

5. Ibid. (abridged form).

LELAND H. WILLIAMS
Triangle Universities
Computation Center

24

A Functioning Computer Network for Higher Education in North Carolina*

Introduction

Currently there is a great deal of talk concerning computer networks, so much that the solid achievements in the area sometimes tend to be overlooked. It should be understood, then, that this paper deals primarily with achievements. Only the last section, which is clearly labeled, deals with plans for the future.

In the terminology of Peterson and Veit,[1] TUCC is essentially a centralized, homogeneous network comprising a central service node (IBM 370/165), three primary job source nodes (IBM 360/75, IBM 360/40, IBM 360/40), twenty-three secondary job source nodes (leased-line Data 100's, UCC 1200's, IBM 1130's, IBM 2780's, and leased- and dial-line IBM 2770's), and about 125 tertiary job source nodes (sixty-four dial or leased lines for Teletype 33 ASR's, IBM 1050's, IBM 2741's, UCC 1035's, and so on) (see Figures 24.1 and 24.2). All source node computers in the network are homogeneous with the central service node and, although they provide local computational service in addition to teleprocessing service, none currently provides nonlocal network computational service. The technology for providing network computational service at the primary source nodes is immediately available, however, and some cautious plans for using this technology are indicated in the last section.

Background

The Triangle Universities Computation Center (TUCC) was established in 1965 as a nonprofit corporation by three major universities in North Carolina: Duke University at Durham, The University of North Carolina at Chapel Hill, and North Carolina State University at Raleigh. Duke is a privately endowed institution and the other two are state-supported. Among them are two medical schools, two engineering schools, 30,000 undergraduate students, 10,000 graduate students, and 3,300 teaching faculty members.

* Reprinted, with permission, from the *Proceedings of 1972 Fall Joint Computer Conference*, AFIPS Press, Montvale, New Jersey, 1972, pp. 899–904.

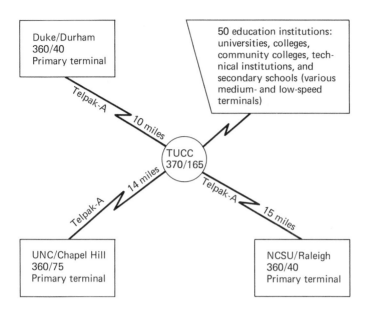

Note: In addition to the primary terminal installation at Duke, UNC, and NCSU, each campus has an array of medium- and low-speed terminals directly connected to TUCC.

Figure 24.1. The TUCC network.

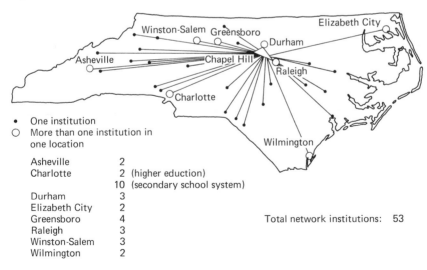

Figure 24.2. Network of institutions served by TUCC/NCECS.

The primary motivation was economic—to give each of the institutions access to more computing power at lower cost than they could provide individually. Initial grants were received from the National Science Foundation and from the North Carolina Board of Science and Technology, in whose Research Triangle Park building TUCC was located. This location represents an important decision, both because of its geographic and political neutrality with respect to all three campuses and because of the value of the Research Triangle Park environment.

The Research Triangle Park is one of the nation's most successful research parks. Located in a wooded tract of 5,200 acres in the small geographic triangle formed by the three universities, the Park in 1972 had 8,500 employees, a payroll of $100 million, and an investment in buildings of $140 million. The Park contains forty buildings that house the research and development facilities of nineteen separate national and international corporations, government agencies, and other institutions.

TUCC pioneered massively shared computing; hence there were many technological, political, and protocol problems to overcome. Successive stages toward solution of these problems have been reported by Brooks, Ferrell, and Gallie,[2] by Freeman and Pearson,[3] and by Davis.[4] This paper will focus on the present success.

Present Status

TUCC supports educational, research, and (to a lesser but growing extent) administrative computing requirements at the three universities, and also at fifty smaller institutions in the state and several research laboratories by means of multispeed communications and computer terminal facilities. TUCC operates a two-megabyte, telecommunications-oriented IBM 370/165 using OS/360-MVT/HASP and supporting a wide variety of terminals (see Figure 24.3). For high-speed communications, there is a 360/75 at Chapel Hill and there are 360/40's at North Carolina State and Duke. The three campus computer centers are completely autonomous. They view TUCC simply as a pipeline through which they get massive additional computing power to service their users.

The present budget of the center is about $1.5 million. The Model 165 became operational on September 1, 1971, replacing a saturated 360/75, which had been running a peak load of 4,200 jobs per day. The life of the Model 75 could have been extended somewhat by the replacement of 2 megabytes of IBM slow core with an equal amount of Ampex

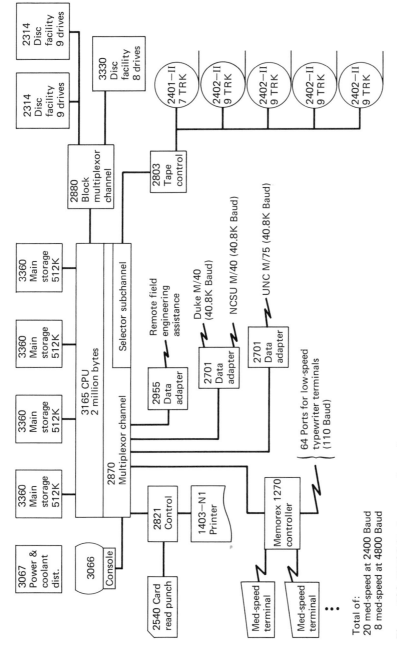

Figure 24.3. TUCC hardware configuration.

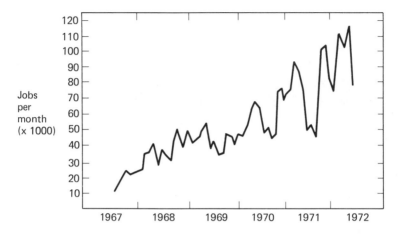

Figure 24.4. TUCC jobs per month, 1967–1972.

slow core. This would have increased the throughput by about 25 percent for a net cost increase of about 8 percent.

TUCC's minimum version of the Model 165 costs only about 8 percent more than the Model 75, and it is expected to do twice as much computing. So far it has processed 6,100 jobs per day without saturation. This included about 3,100 autobatch jobs, 2,550 other batch jobs, and 450 interactive sessions. Of the autobatch jobs, 94 percent were processed with less than 30 minutes' delay (probably 90 percent with less than 15 minutes' delay), and 100 percent with less than 3 hours' delay. Of all jobs, 77 percent were processed with less than 30 minutes' delay, and 99 percent with less than 5 hours' delay. At the present time about 8,000 different individual users are being served directly. The growth of TUCC capability and user needs to this point is illustrated in Figure 24.4.

Services to the TUCC user community include both remote job entry (RJE) and interactive processing. Included in the interactive services are programming systems employing the BASIC, PL/1, and APL languages. TSO is running in experimental mode. Available through RJE is a large array of compilers including FORTRAN IV, PL/1, COBOL, ALGOL, PL/C, WATFIV, and WATBOL. These language facilities coupled with an extensive library of application programs provide the TUCC user community with a dynamic information-processing system supporting a wide variety of academic computing activities.

Advantages

The financial advantage deserves further comment. As a part of the planning process leading to installation of the Model 165, one of the universities concluded that it would cost about $19,000 per month more in its hardware and personnel costs to provide all computing services on campus than would continued participation in TUCC. This would represent a 40 percent increase over their present expense for terminal machine, communications, and their share of TUCC expense.

There are other significant advantages. First, there is the sharing of a wide variety of application programs. Once a program is developed at one institution, it can be used anywhere in the network with no difficulty. For proprietary programs, usually only one fee need be paid. A sophisticated TUCC documentation system sustains this activity. Second, there has been a significant impact on the ability of the universities to attract faculty members who need large-scale computing for their research and teaching, and several TUCC staff members have adjunct appointments with the university computer science departments.

A third advantage has been the ability to provide very highly competent systems programmers (and management) for the center. In general, these personnel could not have been attracted to work in the environment of the individual institutions because of salary requirements and because of system sophistication considerations.

North Carolina Educational Computing Service

The North Carolina Board of Higher Education has established an organization known as the North Carolina Educational Computing Service (NCECS). This is the successor of the North Carolina Computer Orientation Project,[5] which began in 1966. NCECS participates in TUCC and provides computer services to public and private educational institutions in North Carolina other than the three founding universities. Presently forty public and private universities, junior colleges, and technical institutes, plus one high school system, are served in this way. NCECS is located with TUCC in the North Carolina Board of Science and Technology building in the Research Triangle Park. This, of course, facilitates communication between TUCC and NCECS, whose statewide users depend upon the TUCC telecommunication system.

NCECS serves as a statewide campus computation center for users, providing technical assistance, information and other related services. In addition, grant support from the NSF has made possible a number of curriculum development activities. NCECS publishes a catalog of available instructional materials, and provides curriculum development services. Its staff offers workshops to promote effective computer use and visits campuses, stimulating faculty to introduce computing into courses in a variety of disciplines. Many of these programs have aroused interest in computing in institutions and departments where there was no interest at all. One major university chemistry department, for example, ordered its first terminal in order to use an NCECS infrared spectral information program in its courses.

The software for NCECS systems is derived from a number of sources in addition to sharing in the communitywide program development described above. Some of it is developed by NCECS staff to meet a specific and known need; some is developed by individual institutions and contributed to the common cause; some of it is found elsewhere and adapted to the system. NCECS is interested in sharing curriculum-oriented software in as broad a way as possible.

Serving small schools in this way is both a proper service for TUCC to perform and is to its own political advantage. The state-supported founding universities, UNC and NCSU, can show the legislature how they are serving much broader educational goals with their computing dollars.

Organization
TUCC is successful not only because of its technical capabilities but also because of the careful attention given to administrative protection of the interests of the three founding universities and of the NCECS schools. The mechanism for this protection can, perhaps, best be seen in terms of the wholesaler-retailer concept.[6] TUCC is a wholesaler of computing service; this service consists essentially of computing cycles, an effective operating system, programming languages, some application packages, a documentation service, and management. The TUCC whole-sale service specifically does *not* include typical user services such as debugging and contract programming. Nor does it include user-level billing or curriculum development. These services are provided for their constituents by the campus computation centers and NCECS, which are the retailers for the TUCC network (see Figure 24.5).

The wholesaler-retailer concept can also be seen in the financial and service relationships. Each biennium, the founding universities negotiate with each other and with TUCC to establish a minimum financial commitment from each to the net budgeted TUCC costs. Then on an annual basis the founding universities and TUCC negotiate to establish the

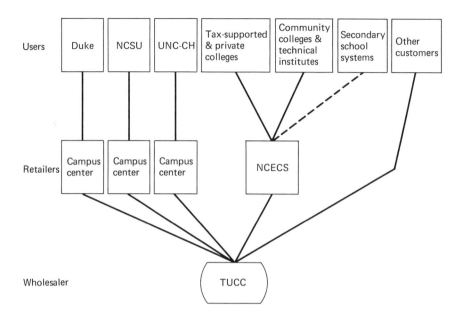

Figure 24.5. TUCC wholesaler-retailer structure.

TUCC machine configuration, each university's computing resource share, and the cost to each university. This negotiation, of course, includes adoption of an operating budget. Computing resource shares are stated as percentages of the total resource each day. These have always been equal for the three founding universities, but this is not necessary. Presently each founding university is allocated 25 percent, the remaining 25 percent being available for NCECS, TUCC systems development, and other users. This resource allocation is administered by a scheduling algorithm that ensures that each group of users has access to its daily share of TUCC computing resources. The algorithm provides an effective trade-off for each category between computing time and turnaround time; that is, at any given time the group with the least use that day will have job selection preference.

The scheduling algorithm also allows each founding university and NCECS to define and administer quite flexible, independent priority schemes. Thus, the algorithm effectively defines independent submachines for the retailers, providing them with the same kind of assurance that they can take care of their users' needs as would be the case with totally

independent facilities. In addition, the founding university retailers
have a bonus because the algorithm defaults unused resources from other
categories, including themselves, to one or more of them according to
demand. This is particularly advantageous when their peak demands do
not coincide. This flexibility of resource use is a major advantage that
accrues to the retailers in a network like TUCC.

The recent installation of the old TUCC Model 75 at UNC deserves
some comment at this point because it represents a good example of the
TUCC organization in action. UNC has renewed a biennial agreement
with its partners that calls essentially for continued equal sharing in
the use of and payment for TUCC computing resources. Such equality is
possible in our network precisely because each campus is free to
supplement as required at home. The UNC Model 75 is a very modest
version of the prior TUCC Model 75. It has 256K of fast core and one
megabyte of slow core, where TUCC had one and two megabytes, respectively. Rental accruals and state government purchase plans combined to make the stripped Model 75 cost UNC less than its previous
Model 50. It provides a 20 percent throughout improvement over the
displaced Model 50. The UNC Model 75 has become the biggest computer terminal in the world!

There are several structural devices that serve to protect the interests of
both the wholesaler and the retailers. At the policy-making level this
protection is afforded by a board of directors appointed by the chan-
cellors of the three founding universities. Typically, each university
allocates its representatives to include (1) business interests, (2) computer
science instructional interests, and (3) the interests of its other computer
users. The university computation center directors sit with the board
whether or not they are members, as do the director of NCECS and the
president of TUCC. A good example of the policy level function of this
board is its determination, based on TUCC management recommendations, of computing service rates for NCECS and other users.

At the operational level there are two important groups, both normally
meeting each month. One is the campus computation center directors'
meeting that also includes the director of NCECS and the president,
the systems manager, and the assistant to the director of TUCC. The systems programmers' meeting includes representatives of the three universities, NCECS, and TUCC. In addition, each of the universities has
the usual campus computing committees.

Prospects

TUCC continues to provide cost-effective general computing service for its users. Some improvements that can be foreseen include

1. A wider variety of interactive services to be made available through TSO.
2. An increased service both for instructional and administrative computing for the other institutions of higher education in North Carolina.
3. Additional economies for some of the three founding universities through increasing TUCC support of their administrative data-processing requirements.
4. Development of the network into a multiple-service node network by means of the symmetric HASP-to-HASP software developed at TUCC.
5. Provision (using HASP) for medium-speed terminals to function as message concentrators for low-speed terminals, thus minimizing communication costs.
6. Use of line multiplexors to reduce communication costs.
7. Extension of terminal service to a wider variety of data rates.

Administrative Data Processing

Some further comment can be made on item 3. TUCC has for some time been handling the full range of administrative data processing for two NCECS universities and is beginning to do so for others. The primary reason that this application lags behind instructional applications in NCECS schools is simply that grant support, which stimulated development of the instructional applications, has been absent for administrative applications. But word of the success of the two pioneers has already begun to spread among the others.

With the three larger universities there is a greater reluctance to shift their administrative data processing to TUCC, although Duke has already accomplished this for its student record processing. One problem that must be overcome to complete this evolution and allow these universities to spend administrative computing dollars on the more economic TUCC machine is the administrator's reluctance to give up a machine on which he can exercise direct priority pressure. The present thinking is that this will be accomplished by extending the submachine concept (job scheduling algorithm) described in the previous section so that each founding university may have both a research-instructional submachine and an administrative submachine with unused resources defaulting from either one to the other before defaulting to another category. Of course, the TUCC computing resource will probably have to be increased to accommodate this; the annual negotiation among the founders and TUCC provides a natural way to define any such necessary increase.

Summary

Successful massively shared computing has been demonstrated by the Triangle Universities Computation Center and its participating educational institutions in North Carolina. Some insight has been given into the economic, technological, and political factors involved in the success as well as some measures of the size of the operation. The TUCC organizational structure has been interpreted in terms of a wholesale-retail analogy. The importance of this structure and the software division of the central machine into essentially separate submachines for each retailer cannot be overemphasized.

References

1. J. J. Peterson and S. A. Veit, "Survey of Computer Networks," MITRE Corporation Report MTP-359, MITRE Corporation, McLean, Va., 1971.

2. F. P. Brooks, J. K. Ferrell, and T. M. Gallie, "Organizational, Financial, and Political Aspects of a Three University Computing Center," *Proceedings of the IFIP Congress*, North Holland Publishing Co., Amsterdam, distributed in the U.S.A. by AFIPS Press, Montvale, N.J., 1968, pp. E49–52.

3. D. N. Freeman and R. R. Pearson, "Efficiency vs. Responsiveness in a Multiple-Service Computer Facility," *Proceedings of the 1968 ACM Annual Conference*, Association for Computing Machinery, New York, pp. 24–34b.

4. M. S. Davis, "Economics—Point of View of Designer and Operator," *Proceedings of the Interdisciplinary Conference on Multiple Access Computer Networks*, Department of Electrical Engineering, University of Texas, and MITRE Corporation (distributed only at the Conference, April 20, 1970), pp. 4-1-1 through 4-1-7.

5. L. T. Parker, T. M. Gallie, F. P. Brooks, and J. K. Ferrell, "Introducing Computing to Smaller Colleges and Universities—A Progress Report," *Communications of the ACM*, Vol. 12, 1969, pp. 319–323.

6. D. L. Grobstein and R. P. Uhlig, "A Wholesale Retail Concept for Computer Network Management, *Proceedings of the 1972 Fall Joint Computer Conference*, Vol. 41, Part II, AFIPS Press, Montvale, N.J., 1972, pp. 889–898.

25

Network Management
Report of Workshop 5

JULIUS ARONOFSKY
Southern Methodist University
Faculty Discussion Leader

Contributors:

KAREN AH-MAI
University of Hawaii

JACK ARNOW
Interactive Data Corporation

ERIC AUPPERLE
University of Michigan

KENNETH L. BOWLES
University of California,
San Diego

JAMES C. EMERY
University of Pennsylvania

LYNN HOPEWELL
Network Analysis Corporation

JOHN ISELI
The MITRE Corporation

FREDERICK G. KILGOUR
Ohio College Library Center

HARVEY J. MC MAINS
American Telephone & Telegraph Company

ALBRECHT J. NEUMANN
National Bureau of Standards

ROBERT F. SHARP
University of California,
San Francisco

LELAND H. WILLIAMS
Triangle Universities Computation Center

Introduction

Organizing and running a computer network involving different institutions and decentralized users with separate affiliations and dissimilar interests present special management problems. The workshop formulated a specific action plan as a basis for understanding the management problems and as a means for getting started. This action plan was based on the assumption that a resource-sharing educational network of national scope could emerge, possibly on a voluntary basis, provided the right decision makers were brought together and were provided with relevant background material and a suitable decision-making mechanism. The workshop participants formulated the sequence of steps that would have to be considered and evaluated in order to test whether the proposed plan was viable. They made the assumption that the existing regional networks and specialized information centers can serve as a solid base upon which to build. They made the further conservative assumption that extensive external funding would not be available for the installation and operation of the proposed resource-sharing network and, therefore, that it would make sense to start with the existing regional networks and to propose means for linking them. Beginning the telecommunication connections on a piecemeal basis would simplify the cost-benefit analysis and the initial technical design.

Objectives

The overall objective of national resource sharing is to bring resource-sharing capability to all those who can profit from available resources. These resources must be provided on a reliable, easily accessible, economically viable, enduring basis. Services must be available not only at the large, well-endowed institutions but also at medium and small-sized colleges and schools, and in public and private secondary and perhaps even elementary educational systems. Special services must be made available in regions and localities that are geographically distant from major resource centers.

In discussing resource sharing, it is useful to define certain terms:
1. *Development.* The concepts, criteria, methodologies, procedures, techniques, hardware and software construction, and linkages.
2. *Computational resources.* Main frame processing units, peripheral storage devices, input-output devices, specialized equipment, and operating systems software.

3. *Functional capabilities.* Computer programs, procedures, and ancillary support and utility systems.
4. *Information.* All types of data bases and collections of information.
5. *Expertise.* The knowledge, talents, and accrued experience of all those working in the fields and specialties affected, specifically recognizing and including the organized and controlled feedback, analysis, and evaluation of past and current efforts in related systems, practices, technologies, and disciplines.
6. *Support Services.* Educational and training materials and documentation at all levels; user assistance, resource maintenance, resolution of problems, and response to unforeseen exigencies.
7. *Organization, Communication, and Control.* The effective means of binding together the resources listed here.

Possible goals and objectives for a national science computer network include the following:

1. To make available the collective resources of the educational and research community in the United States (and at a later stage the international scene) to members of that community and to related activities.
2. To achieve a more effective allocation of available resources as well as those which can be shared later.
3. To achieve economies of scale resulting from a broad base of users and from more efficient utilization of resources by load balancing, resource specialization, and avoidance or reduction of redundant efforts.
4. To develop and disseminate a comprehensive inventory of national resources in education.

Catalog of Current Information
Potential organizers, operators, and users of future computer networks have a need for current information on network development. The workshop proposed a two-step cataloging operation. The first step is to classify information about existing resource-sharing computer networks both to help develop a national science network and to stimulate the organization of additional regional resource-sharing systems. The second step is to catalog resources able to be shared on networks. This information would be obtained primarily from regional networks, such as MERIT and TUCC, major universities, disciplinary centers, and user organizations.

The resource-sharing catalog should be organized carefully, with resources summarized as hardware, software, data bases, and human expertise. The catalog should note the availability of similar, specialized, and substitute equipment. It would identify programs of proved dependability and widespread utility. It would list the major scientific data bases in the nation that are open to inquiry, such as MEDLARS, the clearinghouse for census data, and the data of the National Center for Atmospheric Research. It would describe the skills and specialty areas of individuals available for consultation. The primary intent of the catalog would be to give network users access to network experience and knowledge that exists but is often not easily available to them.

Regional Networks
Workshop participants noted the valuable service being performed by the growing number of regional networks. Many institutions not now part of a network still have an interest in joining a regional center or in forming a new one. While discussions go on with respect to national networking, selected institutions should be encouraged to explore the advantages of creating additional regional centers. These regional centers may prefer to function independently for some time into the future before becoming linked to a national network. This would depend on the extent to which the promises of national networking were backed up by actual experience.

Financing
The financial support of many present regional and national networks is provided by government and private agencies who do not necessarily draw direct benefits from these networks themselves. But support for a National Science Network must provide direct and tangible benefits to the participants if they are to pay the costs. The initial planning and solicitation of prospective users must provide a financial strategy that demonstrates these benefits.

Current sources of outside funding may be available to develop the studies and proposals necessary to carry out the first phase of the network development plan. The planning for the network's start-up operation must encourage widened usage through the demonstration of reasonable economy, self-sufficiency, and true value to the users. New investments in networking capability should be cost justified on the basis of a careful risk analysis, and research projects should be financed out of ongoing funds and network revenue on the basis of their potential benefits to the network operation.

Effective financial control could be difficult unless direction of the network were centralized in a single organization. Some reasonable prospect of long-term funding will be required.

Initial Participants

The workshop recommended that an effort be mounted to attract the participation of twenty to forty institutions in the formation of a network organization. Since these institutions would be required to commit money and personnel to the enterprise, it would be necessary to communicate effectively with their senior management. A mail solicitation by a small nucleus of highly qualified people hired to spend nearly full time in the organizing activity over a period of months may be indicated in getting started.

The initial participants should be a balanced group of suppliers of computing and information services, members of user organizations, and users of regional networks. Once the initial participants are chosen, they could proceed to organize and arrange financing for the structure needed to implement the network.

Network Organization

The following guidelines might govern the organization and operation of a national computer network:

1. The network will consist of a distributed set of retailers and suppliers, where a supplier is any computing facility offering service to some set of users probably through a retailer facility. There will probably be many more retailers than suppliers, though all will be physically interconnected via an effective communication system.

2. The suppliers generally will be independent administrative entities offering their services as wholesalers to retailers in an open market and on a cash basis. Their rates will be expected to provide full recovery of all costs. Retailers in turn will bill their users at rates they may set independently.

3. The retailer's ability to charge a higher rate than the supplier is based on his providing documentation, counseling, and educational services; arranging for all network administrative services; and offering local services to interactive and batch terminals.

The network should be organized centrally only insofar as is necessary to provide a nationwide facility for the sharing and interchange of computer-based resources. For some unique facilities, such as a national discipline-oriented library, the network would provide users a convenient means of access. For computing services, the purpose of the network would be to create a national marketplace whereby suppliers and customers could be brought together. Centrally managing or administering the several facilities supplying services to the network could cause excessive political problems that might be avoided by allowing user feedback within the marketplace and continuous competitive control of server enterprises to work out a matching of resources to user requirements over the entire country.

Users will in many cases gain access to supplier's services through the facilities of a small retail outlet. This outlet might consist of a small computer and its management, concentrating on the input-output devices needed to make efficient use of a large computer at a distance. In other cases, large computer centers might fill both wholesale and retail roles for distinct groups of customers.

The central network management is conceived as having responsibility to operate the communications facilities. It would also enter into contractual relationships with suppliers and user organizations to assure compliance with established conventions and policies regarding such issues as prices and long-term operational stability. These policies would be established and maintained by the controlling board of the network in consultation with network participants.

Operating Policies and Price Mechanisms
To protect both users and suppliers against disruptions introduced by third parties, it will be necessary to establish ground rules and operating policies for participation in the network. The formulation of these policies should be a responsibility of the controlling board of the network.

In general, computing services exchanged over the network will be accompanied by payments from users to suppliers, frequently via the intermediary of a retail outlet. The network might operate a clearinghouse to simplify this process. Both the wholesale suppliers and the retail outlets will have to recover most of their costs from these payments. Clearly, pricing policies will be required covering all participants to prevent short-term price wars or other tactics whereby one supplier may force another out of business by drawing from a short-term subsidy. On the other hand, competition favors the users and should be fostered as long as it does not endanger long-term stability of services.

Disputes regarding pricing policies should be resolved by the governing board of the network. The board can make use of the contractual relationship of the network with each supplier as an enforcement tool.

Selecting Network Technology

The workshop believed that the technical means to create national networks already exist. There are a number of potential topologies, signaling techniques, and hardware configurations that could be used in creating the desired marketplace and distribution mechanism. The task of the organizing group is to select from among the vendors and components of existing networks, such as ARPANET, TYMNET, and the General Electric Timesharing Network, those that could best achieve the objective of bringing a national resource-sharing network into operation. A significant amount of research and development will still be required regarding higher-level communication and user products for remote job entry, file transfer, common network control language, graphics, and so on.

Education

A substantial effort must be devoted to education and training for a broadly based network to succeed. This effort, along with the providing of suitable publications, is among the most important functions that must be assumed by the management of a national science network. A broad program of education would have to reach not only those directly associated with the network but also institutional administrators including presidents, vice-presidents, librarians, computer center directors, department chairmen, and information system managers. A main topic of interest for all these people would be interinstitutional cooperation by means of networking.

A larger educational program would also have to be mounted for network users and contributors, including faculty members, research workers, librarians, systems programmers, and information analysts. Perhaps the largest educational effort of all would be training programs to instruct user groups and their members in the use of data banks and applications software packages.

ROBERT GILLESPIE
University of Washington

26

University Relations with Networks: Forcing Functions and Forces

Author's Framework

The University of Washington, with a student population of 33,000, gives a primary emphasis to research (fifth in the nation in federal grant and contract awards in 1971). The Computer Center is the usual hard-pressed organization, caught balancing continuously growing educational and research needs against fixed budgets. A Control Data Corporation 6400 is used to produce 600,000 jobs a year for academic computing at an operating cost of about $40 per student per year.

The organization of computing at the university breaks computing into administrative and academic areas, with administrative system analysis and programming separated from the Computer Center (which reports to the Vice-President for Research). Other computing functions include data processing for the hospitals operated by the university and more than fifty small to medium-sized computers used in research projects.

Coordination of higher education in the state of Washington is carried out by a Council of Presidents. This structure has affected approaches to resource sharing in computing at a state level. Rather than pressure developing from a chancellor's office, as happens in many states, a voluntary effort was initiated in 1972. A committee of computer center directors in the state was formed to explore approaches to computer resource sharing and networks under the sponsorship of the council. Of course, part of the stimulus for cooperation has been concern that hasty action might be initiated independent of the computer centers and schools.

Washington State University received a grant from the National Science Foundation to establish network facilities in the eastern part of the state, which is sharply cut off from the western part by the mountains. In the west, the Computer Center at the University of Washington has provided limited terminal service to other installations but has not devoted a significant amount of resources to that task (since its own needs exceed its general capacity).

The major policies that affect decision making and the outlook for resource sharing are the following:

1. Operation of the Computer Center as a self-sustaining organization (forcing it to depend on the sale of grant and contract services, and so on, to balance its budget) on a two-year basis with no revolving fund for capital acquisitions.

2. Allocation of the university's share of support of the Computer Center to the colleges, through the Provost, through funds ("green stamps") that can be used only for computer services.

A major assumption made by the Computer Center management (but not yet clearly reflected in the university administration) is that it should be isolated as much as possible from the establishment of the requirements and allocation of computer resources. The separation would allow a clearer management direction and measure of performance for the Center and a clearer approach to the general question of satisfying the university's computing requirements, where the Computer Center is clearly *one* of several sources meeting computing needs. In simple terms, we need to try to avoid asking the barber if we need a haircut.

Major Issues and Forcing Functions
The forces affecting university computing centers can be categorized as those pushing the computer centers into resource sharing (positive) and those pushing away from sharing (negative).

Positive
The lack of resources (mainly money) to meet the expanding computing needs provides the strongest stimulus for resource sharing at this time. Funding limits in all areas (state, university, federal) push the examination of computer resource sharing. The successes of networks in some environments lead to the assumption that this is a feasible approach. But the problem of resource sharing includes defining computer services.

State and university policies with regard to financing provide a strong impetus to sharing. Many centers, operated as self-sustaining enterprises, are unable to obtain the capital they need to expand or improve their services, even though they are in financial balance. Starved for capital, their services (and incomes) are declining. Thus, the alternative of sharing either to add income or to avoid capital investment is quite attractive.

Negative
One of the major problems involved with resource sharing is not, as is

often expressed, the truculence and empire-building nature of computer center directors (read: natural entrepreneurs). The problem is the conflict between the role of providing needed computing services and the role of the administrators, who must balance the budget. Few networks involving multiple major nodes that buy and sell services to one another have yet evolved satisfactory solutions to the problems of budgets, planning, priorities, and so on. Generalization from the success of star networks is dangerous, as the resource allocation and control problems change drastically.

Another factor is that the introduction of resource sharing in a declining budget period means changing existing allocations. It is always easy to expand services in a growing budget. As most computer centers have many of their costs fixed, additional funding must come from reductions in people and existing service levels—not an easy task. This, rather than truculence, explains the hesitation to jump into resource sharing.

Then too, while the concept of a free market for services—where prices are used to drive out marginal producers—is attractive, this concept simplifies the problems of pricing services. One major hidden assumption that makes a free market work is that accounting practices are constant so that prices reflect value fairly. Some universities, however, capitalize their equipment and recover only their operating costs. Their presence on a network would result in unfairly attracting use rather than effectively driving out inefficiency. And certain government installations are heavily subsidized, providing artificially low prices. The impact of the loading drastically affects pricing. Few studies have been made about the impact and development of pricing policies in computing centers.

Communication costs still are a major factor in evaluating network trades. And with minicomputer costs still dropping, care must be taken in calculating cost alternatives for computer services. The ARPA network is not yet a viable commercial installation.

The concentration on computer resources causes us to overlook the total cost of computing services—consulting, libraries, teaching—that may not be affected at all by resource sharing. There still will be costs, and they need to be carefully identified.

The problems of planning a single university's computing requirements are complex; achieving agreement among several, without establishing separate organizational relationships, is almost impossible because of the *natural* conflicts in goals, timing, and needs. Few universities and colleges have yet found ways of relating their needs for computing resources to their goals in quantifiable ways (cost per student, number of courses, and so on). Eventually, cost formulas will be used for computing just as they are for libraries. It will be important that these formulas be developed by the people involved rather than imposed.

Another way of viewing the problem of sharing is to consider other efforts at university resource sharing: television, libraries, special facilities, and the like. How many of these efforts led to any success? Why? What were the essential elements?

Approaches to Action

In our state we established a computer network study where the members of the committee were computer center directors at state institutions and representatives of the state data-processing authority. After nine months of conferences, committee meetings, and effort, we produce a very general report. What we all learned from the study was not how to organize the network but how hard it was to reach agreement on questions such as: Who should manage network services? How should services be identified? How can priorities be established? Who will fund any steps? Does anyone agree which other networks have been successes and why? What should a plan of action be?

It was clear that it would be difficult to achieve agreement on a plan that might shift resources from one institution to another. The experience of writing the report was, in a way, a model of running the network—and the model showed up the problems.

The questions we are struggling with are the general management questions associated with centralized versus decentralized service functions. We need to review in detail the factors of funding, accountability, control, planning, and allocation in a way that separates the computer technology and complexities from the basic management problems.

Recommendations

I would recommend attacking the problems of networks and university relationships by studying these questions:

1. Do we agree that computer resource sharing is in the best interests of the university? Do university presidents want to give away their responsibilities for resource control? What are the long-term effects?
2. Would resource sharing be stimulated by organizational changes within the university to separate identification of computer resources needs from the management of computer centers?

3. What methods for priority and allocation are workable with computer networks?
4. What common fiscal practices or standards should be established for administration of networks and centers?
5. What planning methods and measures of effectiveness should be used to judge computing resource sharing?

27

Institutional Relations
Report of Workshop 6

WILLIAM F. MASSY
Stanford University
Faculty Discussion Leader

Contributors:

W. H. BRUNING
University of Nebraska

JAMES L. CARMON
University of Georgia

WALTER FREIBERGER
Brown University

ROBERT GILLESPIE
University of Washington

CHADWICK J. HABERSTROH
University of
Wisconsin—Milwaukee

T. W. HILDEBRANDT
National Center for
Atmospheric Research

WILLIAM B. KEHL
University of California,
Los Angeles

BARBARA MEDINA
Herbert H. Lehman College of the
City University of New York

BEN MITTMAN
Northwestern University

DONALD R. SHURTLEFF
University of Missouri

VINCENT H. SWOYER
University of Rochester

Introduction

This workshop discussed networking from the point of view of college or university administrations as opposed to that of an individual scientist or other user. Two main themes were considered. First, joining a network would imply obvious changes in the role and organization of the university's computing center. These changes will be seen as advantages by some and will be resisted by others. Second, some implications of networking are of special interest to the administration of a university. These may be favorable or unfavorable, depending on the specific circumstances.

This chapter will discuss four general issues: (1) why universities might be interested in networking, (2) specific questions that an administration would want to answer before proceeding, (3) implications of networking for the organization of the university's computer center, and (4) questions about networking that will be of particular concern to top university administrators.

The following section will describe some of the advantages and disadvantages of networking as it affects (1) general-purpose computing, (2) specialized, discipline-oriented computing (3) sharing of information services and data bases, and (4) administrative data processing.

General Issues

Reasons Cited for University Interest in Networking

The reasons why a college or university or its constituent parts may be interested in networking are varied and complex. Different parts of the same university may support (or oppose) participation in a network for quite different and even contradictory reasons. These differences stem in part from different values held by individuals and groups involved, but there is ample evidence that a good deal of confusion exists about what networks are and what changes they are likely to produce.

The arguments in favor of an institution's joining a network can be grouped into three broad classes. (This discussion ignores more personalistic reasons like empire building.)

1. Advancement of science and technology pertaining to networking. Large-scale networks of the kind being considered have never been built, and a value is seen in studying them and pioneering in their development. Problems exist in communications and computing technology, software development, and a myriad of organizational, political, and economic areas.

2. Networks are perceived as offering increased effectiveness in meeting the computing needs of faculty, staff, and students. For example, membership in an appropriate network could allow physicists on a given campus to obtain specialized large-scale computing that could never be economical on that campus. The same applies to those wanting special languages (for example, APL) and data bases.

3. Networks are sometimes seen as increasing the efficiency of computing in colleges and universities. This is most easily translated into reducing the cost of meeting the present computing load.

Of the three classes of objectives, the efficiency goal is likely to be the most disruptive in terms of established university relationships and organizational structures. This is because it implies a substitution of the network's resources for those presently available on campus. The "scientific advance" and "effectiveness" arguments are likely to command a broad base of support because they involve adding to the range of available computing resources rather than substituting one kind for another. Only when the costs and benefits of computing are balanced against those of other academic program inputs (for example, numbers of faculty) or entirely different resource uses (which might be considered by a state legislature) is the goal of efficiency likely to receive much support.

Generally speaking, those who favor the efficiency arguments for networking tend to consider cost-benefit ratios—for example, dollar cost per central processor cycle. Those who advocate this point of view rely on the assumption that economies of scale in computing and central system support services are sufficient to offset increased costs due to signal transmission and interpersonal communication. The possibility of load balancing across time zones or among institutions with different academic calendars is also attractive. Unfortunately, however, data for testing the validity of these assumptions do not yet exist, or at any rate have not been analyzed and reported.

Some of the implications of effectiveness goals run counter to those of efficiency. The classic example in on-campus computing is where the memory size of the machine is dictated by the needs of a few large users yet must be reflected in the rates paid by all users. The network counterpart occurs when the costs of joining the network for the benefit of a particular class of users must be added to those of providing regular on-campus service. Problems may also arise because there is more than one definition of effectiveness in a given situation. For example, a discipline-oriented network might increase the communication among scientists in that discipline across the country while at the same time reducing their communication with scientists in related disciplines at their own institution.

These considerations demonstrate that it is essential for an institution to examine *why* it is or is not interested in joining a computer network. It must decide how it wants to weigh the pros and cons within each class of issues. Academic administrators will recognize that coalitions of user constituencies will form to back or oppose a given network proposal. However, unless sheer political clout is the ultimate basis for choosing among alternatives, it is imperative that the probable effects of networking be well understood and broken into component parts so that a rational compromise can be sought.

Questions That Must Be Asked and Answered
The basic advantage of being a node on a computer network is the flexibility it affords in deciding whether a given service or type of computing should be provided locally or purchased from the network. A university would be able to satisfy extreme demands (large CPU-bound programs, access to a large data base, and so on) and provide unusual or unique services by purchasing from the network. This would allow a university to avoid dedicating an inordinate amount of the local resources to a few users.

If an unusual or unique service became extremely popular with local users, a point would be reached where it would be more economical to provide it locally than to purchase it from the network. This would allow a university to explore new services with a minimum of risk and investment and to defer resource or capital investments to meet peak or changing computing loads. However, participation in a network raises some difficult questions, some of which are administrative and organizational and some of which involve funding and cash flow.

Administrative and Organizational Questions
1. A university administration must recognize that independent of the type of network organization (for example, consortium or corporation), the network must be granted functional authority if it is to be operationally effective. The granting of functional authority implies relinquishing a degree of line authority over university computer operations. Therefore, the details of network organization and the definition of functional authority must be clearly established.

2. Since in a network situation a university is dealing with other institutions, a clearly defined, rational process for planning and for allocating funds and resources must be established.

3. The question of depending on others for a major research and instructional resource must be addressed.

4. The organization and nature of the computer center may have to be revised to participate effectively in a computer network. The question of whether a reorganization would result in the loss of computer professionals and thus reduce the opportunity for important kinds of development activity must be considered (it is discussed more fully in the next section).

5. In general, the administration must address policy matters concerning amortization of computer equipment, computer equipment as capital expenditures, and possible conflicts with established university or state policies.

Funding and Cash Flow

1. Unless a workable balance-of-payments system can be established, a university will have to purchase services from the network with real dollars. This raises the question of the impact and implications of cash flow (both university and grant monies) out of the university.

2. Unless a university has access to additional funds and is willing to spend them for computer services, the question of making money available to buy network services must be addressed. Will funds be made available by (a) simply appropriating additional funds, (b) reducing expenditures on locally owned or leased computer equipment, (c) reducing the number of computer professionals on the payroll, or (d) diverting funds from other programs?

If reduction in hardware or staff is contemplated to make the necessary funds available, questions like these must be addressed: (a) Is the network sufficiently reliable and stable so that the university can confidently reduce or eliminate its internal competence and hardware? (b) Will the users have the necessary access to computer professionals? (c) Will the scarcity or absence of computer professionals endanger significant research projects?

The general feeling at the workshop was that a university with an established computer center must carefully examine the proposed network arrangement and thoroughly evaluate the impact of network membership on academic programs, research and development, the system for delivering user services, and capital expenditures versus recurring costs. The questions raised do not pertain to a small college that might be considering network use but to a university with an established computer center.

The Impact of Networking on University Computing Centers
Members of the workshop believed strongly that university involvement in a network would produce fundamental changes in the orientation and operation of campus computing centers. They felt that when computer center managers support networking it is usually on the assumption that their institution will be a server on the network rather than the user. Still, the group felt that a network should not be viewed as "haves serving have-nots." There is likely to be a redistribution of computing among servers on a network, especially if economies of scale are important.

A given computer center may also support networking because it provides a way to serve particularly demanding groups of specialized users without reducing the efficiency of the center's overall operations. Participation in a network could also have some useful political side effects for the computer center director: "prickly" users could be referred elsewhere for service.

The discussions in the workshop, and indeed in the seminar as a whole, showed clearly that many computer center directors support the idea of networking—and for the right reasons. Yet a good deal of caution was expressed regarding the problems that will occur if large-scale networks lead to pressure to reduce on-campus computing capacity. The forces at work can be better understood if we digress slightly to view the organization structure of a university computing center. The following analysis by Robert Gillespie is germane.[1]

There are three major management tasks associated with the direction of a computer center:
1. Needs: What are the computer service requirements of the users?
2. Resources: How does the center get the resources to meet the requirements?
3. Allocation: How does the computer center allocate computing services to the users?
These areas require the major amount of the director's time at a computer center. These will provide the basis for a model of a computer center as a bureau.

When I use the word bureau I do not mean to use it in a pejorative way. Of course most people use the words bureau and bureaucracy with disdain (and a vision of red tape). The definition that I use for bureau is derived from Anthony Downs's definition in his book *Inside Bureaucracy*. Downs's book is a serious, sober (but amusing) study of the general principles of bureaus and bureaucracies. He successfully captures the structures, processes, and people that make bureaus function. The book allows those with experience to exclaim over his insights (because they match their own), but it would be tedious and abstract to fledglings.

According to Downs, two major characteristics help identify a bureau:
1. The output (since it may involve services rather than goods) is not measurable in a simple way.
2. The market for the services is not one that functions on prices alone (e.g., social welfare).

The success of a business can be measured by profit and loss, whereas measuring the value of the output of the Forest Service, for instance, is complex. And a bureau could exist within a business—particularly where there are services provided. Computer centers, accounting departments, social welfare departments, and the like are all bureaus. While you can ask the count expenditures of a bureau, you can't easily measure the value of functions and services. Another general characteristic of a bureau is that its services are not adjusted by a market. That is, the services provided expand until the bureau runs out of resources. People acquiring the services don't usually pay money for them, as in a business.

Downs observed a number of steps in the growth of a bureau. Ordinarily it starts through the efforts of a zealot who sees the need for a new set of functions and who has the energy and commitment to overcome the basic reluctance to change of existing bureaus (try time sharing or computer-aided instruction). As the bureau grows, the need for the zealot diminishes as orderly procedures and planning become necessary to control the larger and larger resources. Thus the classes of people needed at the different stages change. This is equivalent to the typical growth cycle of business—from the original idea and aggressive decision-making founder to the large, ponderous bureaucracy.

Bureaus function in an environment where competition for resources (between bureaus) is a major effort. Since the output and demands for the services are not simple to measure, acquisition of resources requires careful strategy. Relations between the bureaus and their users and other bureaus take on a significance greater than just information transfer. The transfer methods themselves contribute to the strategies. Informal information transfers are important to the functioning of bureaus.

While we can count the number of jobs run in a computer center, this is only one measure of the output. Some of the services include computer jobs; input preparations; manuals, newsletters, and technical information; classes and instruction; reports on use; libraries; special equipment for plotting and other purposes; consultation; and planning information. Since there is no simple measure of the value and adequacy of the computer center, it satisfies the first part of a definition of the bureau.

Determining a proper allocation of computer center resources among the service demands is one of the major management tasks. And, while most centers operate with dollar charges for computer time, that does not provide a market adjustment for all services. Computer center allocation funds ("funny money") can't be diverted to alternative areas such as chairs, tables, or faculty members.

Thus the center is a bureau operating among a number of other bureaus, competing for resources (more money for the library versus the computer center), struggling to identify the needs and display the value of the services to the end objectives of the school. Of course counting the complaint level is a useful measure. "If there are no complaints you're doing too good a job."

The fact that the computer center is not subject to market forces is fundamental to an understanding of the difficulties of developing and adopting computer networks. Part of this can be accounted for by the well-known problems of inducing change in a bureaucratic organization.

Also significant is the fact that a distributive network will put a given computation center in direct competition with other centers. (This problem does not exist in nearly so direct a form when a "tight" network is developed and run by a consortium.) Once communication has been established in a distributive network, it is feasible for users to shift their demand to centers where the cost-effectiveness ratio for their kind of computing is most advantageous. This migration may occur among users on the center's own campus, and it certainly is a potential danger with respect to those customers being imported from elsewhere on the network.

The existence of a large-scale regional or national network could change campus computing from what is basically a natural monopoly to one of the more economically perfect markets in the country. The technology for achieving a high degree of competition already exists, or will in the near future. Whether such competition occurs will depend on technology and systems capability and the organizational forms and policies of the networks—for example, who owns the host computers, freedom of entry and exit, and price regulation. To the extent that significant competition does occur, it will represent a change of major significance for universities.

There are a great many questions about computer center adaptation to networking. The following partial list and discussion was developed prior to the workshop by one of its members, himself a computer center director.

1. Has the computing center become a useful campus institution rather than just a provider of an off-the-shelf product? If so, how will a network or a consortium affect its function?

Most campus institutions like the library have a long history of evolution. The role of computing facilities in the academic program is still not clearly defined and is evolving. We have academic senates, educational policy committees, department visiting committees, and educational program committees that are familiar with our other campus institutions such as the library. But they have only recently taken cognizance of the computing center and the academic issues related to its use.

If a simplistic view is taken, that computing services are an off-the-shelf product to be purchased at its most economical source, these institutional questions will not be solved. What is necessary is a stability that a computing center, considered as an institution as opposed to a machine, can provide. For example, can an academic department in planning its courses count on a well-supported APL service for a long enough period of time to build an academic program around its use? Within their own campus, faculty can get such a long-term commitment. Computing centers are just beginning to solve these problems, interacting closely with their faculty.

It is only with the third generation of computers that centers have become both fiscally and educationally responsible. Most centers are now able to provide a reasonably good batch service with ease of access for students. In time sharing there still are many problems in data management, file systems, and graphics. If the computing center is a recognized and accepted campus institution, these key academic questions can be answered in an evolutionary manner through continued interaction between the center and the faculty in the context of the campus goals. Will this be lost in a consortium or in a network?

2. Computing centers in the larger universities today provide more than pure numerical computational power. They provide file systems with shared files and data security, a variety of processors and languages and of data bases, and a variety of interactive services. They are much more a sociological phenomenon—an environment in which students and faculty work and have access to the same programs and data files, to debugging tools peculiar to their experiences and changing needs. Does your campus computing center provide such an environment? If so, how does it make its decisions? What will happen to this aspect in a consortium or in a network?

As a sociological phenomenon, the availability of computers brings up many campus issues because of competing needs. These involve the allocation of resources. Many of the characteristics of the computer system the faculty want are not cost effective from an economic viewpoint, and no service bureau would provide them. But they may be cost effective in a total educational sense, and this warrants a different value scheme. The computing center can cater to these local campus needs and changes in academic programs. But the development of exclusive specialization by centers, as sometimes postulated for networks, into batch number crunchers, student time-sharing centers, discipline-oriented centers, and so on, limits the breadth of the local computing environment to users. Moreover, today's computer systems do not require such specialized centers. A third-generation computer system will provide a better balance if it handles a mixture of services including batch and time sharing, small student jobs and large research tasks—both numerical and data manipulation tasks. This balance is something that must be well planned. Unstable changes in these commitments could seriously degrade this system and turn a responsive service into a very poor one. There must be very close working relationships with the users in the management of such a computer system.

3. What are the different management considerations of a consortium and a network? The effect of a consortium is to achieve a certain critical mass. An understanding of the independence of size of the computer from the complexity of the management structure must be weighed. For example, Ohio State's computing center, which has a simple management structure, is larger and carries a greater and more varied load—and, incidentally, a higher budget—than the Triangle Universities Computation Center complex. But, only through accepting the complex problems of a consortium could three North Carolina universities solve their problem of achieving a critical mass. The effect of a network is to make available resources you do not choose to provide locally regardless of the size of your computing center, and perhaps in the case of the larger centers to provide services to others to make it easier to make such decisions.

The management of networks is considerably simpler than the management of a consortium. Pricing on a network is done by each center individually. The flow of work becomes natural, to fulfill a need. In a consortium, the flow of work apparently has to be artificially forced. Loads have to be artificially balanced. Pricing is an experiment, and the stability of the computing center is at stake. Long-term commitments as in a consortium are not a problem for networks, and the centers are not locked into management constraints or agreements on improvements or evolutionary development. Decisions remain local to the individual campus in a network environment. Line responsibility for computing is clearly defined in parallel to the academic program. The reward system for good performance is well established. In a consortium you must give up this autonomy and local campus value schemes in the interest of achieving a critical mass. Usually this decision is for economic reasons, where only the computer costs are considered and the management costs forgotten.

4. Can you muster the technical capability to participate in a network? Surprisingly, it takes more technical capability and resources to participate in a network than to operate a stand-alone computing center. You must be able to deal with the complexities of data communications and, in the case of a heterogeneous network like ARPA, a variety of protocols. Users must have consultants on campus who are better qualified in order to be familiar with the remote services than they need be as local consultants. This is because there is always someone available, perhaps another user, who can help a local problem. You don't have this local group interaction as easily for services provided remotely. You must have systems programmers capable of dealing with the relatively new and technical problems of making modifications you need

locally if your service is to be responsive to your campus. You must have better documentation and a training program. These are significant costs.

The Administration's Point of View

We have already noted that a university's central administration must balance the needs of a number of different constituencies. These include academic users such as members of the faculty and student body, sponsors of research and instructional projects, administrative data-processing and Management Information Systems groups, and the computation center as an institutional force. All these must be balanced with respect to the institution's overall values in both the long and short run. For example, proposals pertaining to the types, levels, and reliability of service in support of instruction and research goals must be traded off against other demands on financial resources.

All the questions cited earlier in this chapter are, of course, germane to the administration viewpoint. Additional questions that reflect this interest are the following:

1. Networking offers the possibility of changing computing from a heavily fixed cost to a largely variable cost operation. This is probably an advantage for the institution as a whole, since it allows closer matching of the need to the supply of computer services. (A "computing lobby" in the university may see this aim as threatening to reduce the commitment to computing over time, but this conclusion is certainly not necessary.) While paying for computing on a variable cost basis will complicate forecasting procedures and tend to generate budget variances, any desired degree of stability can be obtained by the establishment of forward contracts or accounting reserves.

Perhaps the overriding advantage of putting campus computing on a variable cost basis is that it would no longer be necessary to have tight controls on the ability of people to go outside with their computing dollar. There might still be a need to coordinate computing to take advantage of economies of scale reflected in external pricing agreements, but with reasonable management it should be possible to maintain a much higher degree of flexibility than would be possible without a network. In principle, it would even be possible for the university to distribute its own computing funds to students and faculty and allow them to shop around on the network or elsewhere. Nor is there any logical reason why such funds could not be freely exchangeable in other markets—for example, using them to buy the time of research assistants instead of computing or vice versa. None of these arrangements are possible when the institution's annual computing budget must be committed in advance to the operation of an established center and as much sponsored support funneled into it as possible.

2. If a campus retains some equipment whose service it intends to market locally or on the network, will this service be competing with subsidized systems elsewhere on the network? The existence of highly subsidized national discipline-oriented centers could pose a significant problem for campus computing. It might be difficult to limit the use of such centers to a particular type of work, and thus they might compete with general campus computing at prices that are below real costs of operation and equipment depreciation. Reducing the "friction of space" by hooking these centers together on a network would exacerbate these problems. A university's administration will want to assess this effect on campus computing. In addition to the obvious financial questions, the administrators will be concerned about whether the availability of such services will tend to distort their institution's academic priorities.

3. If a network involving the availability of discipline-oriented computing centers and effective data-base sharing should come into existence, could a major university afford not to participate? It seems likely that access to these centers through the network would be a prerequisite for maintaining a high-quality academic program in that discipline. Without such access it would be difficult to hold faculty and to get top graduate students. This point will be a powerful force for further extension of the network.

4. Even though participation in a network may be necessary to maintain a strong program in a discipline with specialized national computing, it is possible that faculty will see exclusive reliance on network resources as a sign of weakness in their university's program. There will be considerable jockeying for position within each discipline among the major universities. Administrators will need to decide early whether or not they will compete in a given area and be prepared to back a positive position with substantial resources. If an institution chooses not to try to become a steeple of excellence in computing in a given discipline but rather to rely on the resources of the network, it will need to take special pains to convince critical faculty members that they should not migrate.

5. It is possible that the spread of networks will reduce the loyalty of faculty members to their individual institutions in favor of the "invisible college" of colleagues in their disciplines. This process has already

proceeded a good distance in a number of fields, partly as a result of patterns of federal research funding. University administrators will need to consider how to cope with this phenomenon.

Special Considerations Pertaining to Types of Computing

Almost every college or university campus requires or will require a general-purpose computing service. The nature of this requirement varies with each institution. The objective for satisfying these needs must be the availability of service, not necessarily the operation or preservation of a computer center as we now know it.

General Computing for Instruction and Research

Computer service will be obtainable from several kinds of sources, of which a network is but one. For many applications the present type of local computer center will be the best solution. One of these may be batch educational applications, where most university centers presently provide excellent service at low cost. Other applications are most reasonably suited to the minicomputer market. Minicomputers frequently have the advantage of low cost geared to a specific problem area, and because of their relatively small size they will generally capitalize on new production and design technology earlier than large systems. They may also be acquired through capital rather than operating funds —a distinct advantage under some circumstances. Specific areas well suited to minicomputer technology might include instruction in elementary programming and the handling of certain file applications where control or confidentiality are important.

Clearly, the need exists for institutions to redefine the services required. The emphasis shifts from trying to exploit the local central processor to analyzing the alternative means of acquiring service.

Provision of general-purpose computing service on a campus by means of a network is the most extreme example of reliance on external supply. This alternative will be attractive if economies of scale in computer hardware and systems are sufficient to offset the costs of communication. Assuming that the communication system itself is reliable, it may be possible to improve the reliability of campus service by providing better backup systems. On the other hand, for a major campus to give up its big-computer business is a very big step. The considerations given earlier in this chapter apply with particular force to general-purpose computing.

Discipline-Oriented Centers

The characteristics of discipline-oriented centers are discussed extensively elsewhere in this volume and will not be described here. The workshop cited the following characteristics and organizational problems associated with such centers as well as other kinds of network arrangements:

1. A specialized center can act as wholesaler of its services, with local university centers as retailers.

2. Billing and access functions can be decentralized to the university level.

3. Risks and high fixed costs can be shared over a relatively larger system.

4. Flexibility of marginal resources and the range of options available are increased.

5. Higher technical virtuosity is demanded of university centers.

6. The range of human resources linked to the universities by a network can be increased. The network may include voice communication so that one can call in experts at other facilities.

7. A pricing and central clearing system must be developed to guarantee recovery of costs of all participating nodes, pricing of all network services at or below the cost of the best alternative, and the accumulation of some kind of surplus resources in the network to cover temporary or emergency disequilibrium in payment clearings.

Characteristics identified as applying almost exclusively to discipline-oriented centers were the following:

1. Specialized centers can easily be related administratively to discipline-oriented funding and project selection procedures. They offer a convenient means of efficient administration of subsidies allocated to disciplines and can offer a means of project control, including special procedures to limit access to authorized users in the discipline supported.

2. Network can provide communication links between geographically dispersed or mobile specialists in the disciplines.

3. Networks and specialized centers can make expensive software, data banks, and equipment unique to a discipline available as if they were local computing center services.

Information Services
Information services encompass the full range of activities associated with the creation, maintenance, storage, retrieval, and manipulation of machine-readable data bases. These data bases include bibliographic files, research data files, library files, and the like and consist of both textual and numeric data. Data bases may be very large (millions of records), like Chemical Abstracts Service, or small (hundreds or thou-

sands of records) such as those compiled by individual researchers. Centers providing information services offer storage of the data bases, remote and batch-oriented retrieval and manipulation software, and user services and documentation.

LARGE-SCALE BIBLIOGRAPHIC DATA SERVICES. The information service that addresses the problem of the storage and retrieval of large bibliographic and library files represents a natural activity for resource sharing. Such services require large investments in magnetic tape services and in software development, which in many cases exceed $500,000 in operational costs. These types of services are obvious candidates for both centralized and distributed network access to data and software that would otherwise be inaccessible to many colleges. Even if a central facility were able to mount such data bases, it would frequently find it infeasible to mix retrieval services with other computing requirements. Even when such services are feasible, cooperation between information centers appears to be a logical approach to resource sharing. One center, providing access to information in discipline A, finds it economical to trade services with another center specializing in discipline B.

One further point is important concerning the development of a number of large-scale information centers. A single center is unlikely to be able to provide for the totality of information services required in the United States. Instead, a number of regional or discipline-oriented information centers is likely to evolve.

LARGE-SCALE RESEARCH DATA SERVICES. Many of the same considerations can be applied to the growth of research data banks for weather, satellites, large-scale medical screening, and so on. As yet these banks are not as far along in their development as bibliographic and library service centers. Nevertheless, they will grow. In addition to retrieval software, these centers will develop applications, statistical, and graphical software that will make them candidates for resource sharing.

SMALL-SCALE RESEARCH DATA BASES. One other aspect of information services that needs to be mentioned is the fact that many of these applications need not be massive enough to require immense resources. Many data-base applications involve small research studies with relatively few, although frequently complex, data bases. Examples of these data bases include those for medical research, political science, sociology, and experimental outputs. These data bases can be organized and interrogated through information retrieval packages and can be processed through associated statistical, graphical, and computational packages that are available on central campus computing facilities.

Another mode of use of such data is via minicomputers with some on-line storage. Studies are needed of the effective use of hierarchical storage and hierarchical processors. How can a researcher acquire a portion of a large data base, transfer it to his local minicomputer, and interact with it there? How can data gathered on a minicomputer be integrated into a larger data base on the central computer and on a network machine? What types of new retrieval, processing, and display software are needed?

OUTSTANDING PROBLEMS AND POSSIBLE TRENDS. The following are a few of the technical and administrative problems that remain to be solved in the provision of effective information services:

1. Maintenance and verification of very large data bases.
2. Optimal distribution of data bases and indexes in hierarchically organized storage systems.
3. Development of effective file structures for on-line systems.
4. Achieving an effective balance between on-line and batch processing.
5. Protection of data rights and copyright for purchased data bases.
6. Protection of income bases of the data-base supplier—the wholesale-retail concept.
7. Combining bibliographic systems with document delivery systems, for instance in libraries.

The following trends and developments seem to be emerging:

1. Information services appear to be more easily accepted in an interstate or regional network than other types of computing services. This is due to the special nature and high cost of providing access to many large data bases.
2. Many services are now in existence that offer opportunities for sharing.
3. Libraries are effective models of the sharing of information without monies changing hands.
4. The problems concerning many of these points have been a concern for over two years of the Association of Scientific Dissemination Centers.
5. Networks and resource sharing are now operating on a trial basis.

Information services are the area of application that have been least affected by political and jurisdictional prejudices. It is unlikely that a large degree of duplication of services will evolve. Specialized regional centers will continue to develop to provide data-base, bibliographic, research, census, and other data services. The requirement for massive on-line or off-line storage capability and for specialized software will discourage "going it alone." The variety of application areas—chemistry, biology, engineering, political science, and so on—will dampen the demand for a single supercenter. Instead, regional and specialized centers will develop gradually and will survive if they can provide services effectively from the point of view of both costs and capability.

The growth of these centers will not inhibit the continued application of data-base systems in central computer facilities or in minicomputer systems where applicable. These applications will be mainly for small research data bases.

Administrative Data Processing
The workshop did not feel that large-scale networking has much to offer universities in the area of ADP. Small schools combine this function with a minimum amount of general-purpose academic computing to good advantage. Large schools have sufficient volume to permit efficient in-house operation. Both types have specialized systems and needs and feel a strong need for administrative privacy and self-controlled reliability.

In its deliberations, the workshop team concerned with ADP placed most of its emphasis on the problem of smaller colleges. In particular, it believed that their needs could best be met by local minicenters having the following characteristics:

1. A minicomputer (local "hands on") with a reader-printer, local tape or disc storage, the option to install interactive terminals for educational needs, and sufficient core for most local data processing.
2. Compatibility with a regional network that provides a reliable, stable, long-term supply of bulk computing power, as needed, and file security for those applications that are handled at the network.
3. Separation of the network from overt control by larger institutions to quiet the fears of smaller schools that they will be gobbled up.

Any research effort in regional or national networking should include provisions for software development to handle subnetworks of minicenters and to provide organizational elements to work out procedures for handling those occasional large files that require security. Of most importance is the need for long-term commitment.

Summary
The workshop's consensus was that networking offers the possibility of significant advantages for colleges and universities. These advantages include the provision of better services of both a general and specialized nature, and conceivably the maintenance of the existing service levels at

lower cost. Considerable caution was expressed that these benefits are still only potential. University administrations must not automatically assume that they will be achieved.

Even if the technical and economic advantages discussed in the workshop can be proved to exist, there still will be many barriers to the rapid, widespread adoption of networking. These include the necessity for changing the function of the university computing center, the variety of objectives and perceptions held by the faculty and students in most institutions, and the possibility that the period of transition to networking will involve more rather than less cost.

The workshop concluded that research is urgently needed both on the technical-economic characteristics of networking and on the institutional-organizational changes that will be required of colleges and universities.

Reference

1. Robert Gillespie, "University/Computer Center Interfaces or One More Bureau in a Bureaucracy," paper presented at the August 1970 University of Colorado Seminar for Academic Computer Center Directors.

28

HARRISON SHULL
Indiana University

Resource Sharing
in Theoretical Chemistry

I wish to discuss aspects of two quite different forms of resource sharing
as they involve theoretical chemistry. The first, the Quantum Chemistry
Program Exchange, is an active operation of small magnitude but, I
believe, of considerable value and utility within its sphere of operation.
The second is only in the proposal stage, a National Laboratory for
Computational Chemistry, although it has been widely discussed for
a number of years. The activities surrounding this proposal, and the
discussion of its possible implementation, can perhaps illuminate more
general problems in bringing resource sharing to fruition.

The Quantum Chemistry Program Exchange (QCPE)

The QCPE was inaugurated in 1962, but a long history preceded its
formation. As the accompanying extract of an earlier conference
indicates, there is a natural propensity towards resource sharing in
theoretical chemistry because of the extraordinary length and complexity of the computations involved. This was recognized at an early
stage, and just after the Shelter Island Conference of 1951 a special
meeting was held to discuss the exchange of computed integrals. The
University of Chicago was designated as a communication center, and
some activity in this direction followed. The proposed exchange did
not bear fruit, however, largely as a result of three factors, in my
judgment: the original concept was nonviable, interest in the professional community was insufficient, and there was an insufficient commitment of individuals to the concept.

Despite this failure, the general concept of resource sharing was discussed at each of the fairly frequent meetings of individuals in the field.
The rapid increase in the number of stored program computers changed
the concept from one of exchange of data to one of exchange of
programs, a much more viable concept since it permitted the indi-
viduality of needs for integrals to be satisfied by use of general computational programs. A long discussion was held at the Gordon
Conference on Theoretical Chemistry in 1962. The exchange idea was
supported by the increasing labor of computing programming, the
rapid obsolescence of equipment, and the rapidly increasing numbers
of newcomers to the field. There was general understanding of the

relatively slow rate of progress being made as a result of having to start from scratch with the introduction of each new computer, and there was hope that the exchange of programs would make more effective use of people and machines.

There was, however, little idea as to how to proceed. In that atmosphere, I offered to try an experiment. We offered nineteen programs, most of them developed at Indiana, through the medium of a mimeographed newsletter distributed to people known to be active in the field. Many of these programs were little more than interesting subroutines, but nonetheless they were useful.

QCPE had some important characteristics from the very beginning. We had identified and responded to the needs of a user community. We were not seeking a market for our ideas; rather we were filling a market need. We had a small but interested group willing to act as entrepreneur for the project. We devised an extremely simple operation that was feasible for us and that did not demand too much in effort or cost on the part of the user. The users perceived a benefit that outweighed the potential costs of belonging.

Shortly after we began operation we received support form the Air Force Office of Scientific Research. This support continued until the sudden shifts in defense research support occasioned by the Mansfield amendment in 1970. We received a terminal grant from the National Science Foundation for a three-year period on a declining basis. Essentially this support has provided the risk capital for continuing the QCPE on a paying basis by its membership. We have successfully made the transition to a basis requiring an annual membership fee and payment for services rendered. There are now more than a thousand members and several hundred programs available for exchange.

In summary, the essential ingredients to the development of this small but apparently successful form of resource sharing are (1) a viable idea, (2) entrepreneurs, (3) capital, (4) a workable procedure or operation, and (5) an adequate market for a product of value priced at a level sufficient to maintain the operation.

The National Laboratory for Computational Chemistry

The concept, as far as I am aware, was first set down in a document of mine written in 1966 and distributed fairly widely by 1968. Later in this chapter is my summary, written in 1970, of the proposal. The concept is straightforward:

• Theoretical chemistry has developed new computational tools that

should be of great value in solving problems of chemistry in general.
- The cost of providing this computational tool is beyond the capability of any individual institution other than the federal government in the foreseeable future.
- The potential benefits of providing the resource are seen by the proposers as far exceeding these costs.

The idea was discussed at a conference of about twenty-five theoretical chemists in Bethesda in 1970. After vigorous discussion and presentation of numerous opposing views, the group nevertheless unanimously declared itself in favor of the concept. The report of the conference was reveiwed by the Committee on Science and Public Policy of the National Academy of Sciences. This committee recommended that the idea be explored further in the context of a wider audience less likely to be viewed as self-serving and in the context of a wider set of problems.

A new panel was formed on this basis, and there have again been heated discussions surrounding the viability of the concept. Despite the seeming disparity of viewpoints and ideas, there is apparently evolving from this group a fairly widespread view that the idea is worth funding. The very arguments surrounding the discussions of the potential national resource are, however, instructive in the consideration of the more general problem of how to begin resource sharing.

The main problems seem to result from
- An inadequate or imprecise understanding of the concept. There are many variations on the central theme, each of which has its proponents and, more important, its antagonists. If the central theme is elaborated too much, almost everyone is against some part of it.
- Inadequate understanding of the needs of the user community. It is hard to project the potential value of a resource to the entire community of chemists, both academic and industrial, when that resource does not exist.
- Perception of the national center as conflicting with the financial status as well as the prestige of present institutions. Even when the greater good of the center is admitted, there is a conflict of interest with respect to the financial viability of computing at the home institution. It is a reverse form, in a way, of the "tragedy of the commons."
- Perception of location at the national center as giving a personal advantage to certain individuals over others not so fortunate. This is the equivalent on the personal level of the institutional conflict of interest mentioned above.
- Distrust of the motives of the proponents of the concept.

• Fear that the financial costs of the new center, which are high relative to the current funding of all fundamental research in chemistry, will deplete the resources for chemistry research in other areas. This assumes a fixed amount of money for chemistry. It is another form of protecting the status quo.

• The ease with which one can oppose the unknown compared with the difficulty of being actively in favor of it.

• Chemists' generally inadequate knowledge of the politics of science. They have mostly stood outside the fray and have suffered the financial consequences of this attitude.

Almost any concerted attempt at resource sharing among existing, well-established, politically strong institutions will run into similar problems. They should be carefully delineated and responded to in detail if the ideas are to gain political acceptance. If, however, the central resource can be established quietly without diminution of perceived current resources of individual institutions and their constituents, then the gradual shift to using the shared resource can occur without having to encounter many of these problems. But to do so requires a quite different form of development of the necessary financial and intellectual capital required for the centralized resource.

Summary of Proposal*

We recommend that steps be taken to establish a National Center for Theoretical Chemistry. Such a center should be financed at such a level that it can encompass within it the finest and most extensive computing facilities chosen explicitly for use in this general subject area, with a permanent staff of theoretical chemists and of systems programmers and operators sufficient to maintain its operations at a high level of competence and efficiency.

Functions and Advantages of a National Center for Theoretical Chemistry

We envisage a number of potential advantages from the existence of a National Center and from its exercise of the function assigned to it.

1. The center will bring into being a permanent group of systems programmers and theoretical chemists who, by their combined long-term efforts, will be able to create programming software of a degree of

* Selections from "Discussions of a National Center for Theoretical Chemistry," by Harrison Shull, May 1970.

sophistication considerably greater than any that can now be created by isolated small groups of individuals not working in concert.

2. The center, because of its preoccupation with a single subject matter area, will be able to procure hardware and develop system software that is especially adapted to and hence especially efficient in its chosen field. It may even be able to influence the production of hardware that will especially enhance computational capability in this field.

3. Individual institutional computing center specifications will no longer have to be tied to the most mammoth institutional users from this field because the largest jobs from these users will be farmed out to the national center. Institutional equipment can then be chosen with more direct attention to the remaining users, most of whom have significantly smaller computer needs.

4. Theoretical chemists seeking large computer time for extensive problems will compete with each other more efficiently for time in a national center rather than compete with instructional users or research users from unrelated and noncomparable fields. A concomitant function of a national center and its advisory councils will be to schedule the available time among theoretical chemists to the extent that the desired use exceeds machine capacity. We feel it is self-evident that this can be done more sensibly than when one must compare projects from totally dissimilar areas. Such allocations also will permit sensible judgments to be made on needs for subsequent expansion.

5. Since the cost per computation on large machines is cheaper the larger the machine, the ability to pool financing and to use the largest possible machine in this area has economic advantages on a per computation basis.

6. The advanced combined education in complex systems capability and complex theoretical competence can only be developed to its highest levels by actual experience in such an environment.

7. Such a center can provide expert assistance in transferring programs that are feasible for running on other machines at other locations to those machines. Such a capability will increase the efficiency of utilization of faculty, students, and machines at these other centers and increase the output of useful computations on molecular systems.

8. Such a center can provide a facility for use of nontheoretical chemists who need the results of computations on chemical systems that they are studying by other techniques. As the National Center develops, it will become an increasingly important resource for data on chemical systems

that cannot be obtained more economically by direct experimental measurement.

Background Information for the Proposal

The Development of Theoretical Chemistry
Theoretical chemistry, broadly speaking, is the application of mathematical methods to the understanding, systematization, and prediction of the properties and reactions of chemical systems. Chemistry is a science most closely linked to the development of materials and to all other physical and biological sciences. Consequently, it plays a vital and basic role in the development of all of science and in the relationship of science to the improvement of human health and welfare.

With the discovery of quantum mechanics in 1926, the groundwork was laid for the possible computation of chemical properties and reactivities by purely mathematical techniques. Unfortunately, the mathematical equations were complicated, and only by arduous labor of dedicated individuals were they solved for the simplest chemical systems in order to test their validity.

Their complexity did not, however, prevent the application of the qualitative insight gained by the study of the equations to the systematics of chemistry, and indeed the language and viewpoint of both inorganic and organic chemistry have been altered at a basic level by the introduction, for example, of molecular orbital theory.

The period of development of theoretical chemistry during the 1930s and during the early postwar years was largely dedicated to the production of elementary models, of the further exploration of extensions of the basic theory, and to the development of the methods of statistical mechanics, which treats molecules en masse rather than as individuals. During this period we have gained important understanding of the energy levels of molecules, the interaction of light with molecules, the computation of bulk properties of matter from the properties of the individual species, and so on. The importance of this period may be said to have received due recognition by the award of Nobel prizes in chemistry to Professors L. Pauling and R. S. Mulliken for their contributions to theoretical chemistry.

The Role of Computers in Theoretical Chemistry
The development of the electronic computer has had an enormous influence upon theoretical chemistry. For the first time, the availability

of machines that could perform mathematical operations at tremendous speeds and that could retain in their "memories" interim and final results and the long and complicated sets of instructions indicating in what order these operations should be performed, enabled theoretical chemists to pursue once again the cherished goal of being able to compute properties of chemically interesting systems.

Even with smaller computers (slower as viewed from today's vantage point), this computational approach had such successes as energy determinations in He and H_2 to accuracies better than those obtainable by experimental methods, the prediction of the structure and energy of a known stable system, H_3+, which is undoubtedly of astrophysical interest, the computation of the first accurately known reaction potential surfaces for molecules, the first clear demonstration of the influence of molecular motion on the changes of solids to liquids, and so on.

As the computers have become larger, there has been an incentive to develop systematic programs that can handle the comparative details of large numbers of chemical systems. Especially valuable among these, for example, has been the effort of the Chicago group to produce detailed computations of a large sequence of diatomic molecules. More recently, limited use of the computers of the present has permitted, for example, a determination for the first time of the nature of the $NH_3 + HCl$ interaction and the conditions under which the individual NH_4Cl molecule can be expected to be stable. Complete computations of modest accuracy have now been done on systems with as many as ten or more atoms and perhaps fifty to sixty electrons.

The Outlook for the Future

The present successes of theoretical chemistry are small both in comparison to the present possible and to future predictable successes. The largest computations done so far have not utilized the full capacity of even the present computers; indeed they have frequently been done on computers which are not even now the largest and fastest available. Furthermore, the technological capability of an increase in computer speed of at least a factor of 100 and perhaps a factor of 1,000 (that is, an increase from about 10^5 instructions per second to 10^7 or even 10^8 instructions per second) is already in sight.

Such increased capability means that computer computations can be made upon many chemically interesting systems, upon material of use in chemical applications, in pharmacy, in engineering technology. We believe that there lies immediately before us the hoped-for period in which we can use theoretical computations to guide the hand of man in using more efficiently the material world around him.

But these are merely the gains we can already see in sight. We cannot yet detail the paths for further development of these tools. On the one hand, it is certain that considered care in the construction of computer hardware can make even more difficult computations feasible. And on the other hand, as we gain knowledge from the results of computations about how molecules are combined and react, we can expect to devise new and efficient systems of programs to enable us to handle more complicated molecules by utilizing directly the results of simple ones, much as a synthetic chemist constructs over more complicated molecules from the previous successes of chemical synthesis.

We can summarize the situation by comparing the electronic computer with other experimental tools such as an accelerator for physicists, a telescope for astronomers, or a mass spectrometer for chemists. The results of computations are not theories. They constitute new data about molecules, about reactions, about materials. For such systems the data form a potential source of information vastly more detailed than can be obtained in any other way. Outstanding examples are unstable reaction intermediates which cannot be isolated long enough for study in the laboratory, and molecular excited states that will undoubtedly provide insights into new synthetic techniques and new ways of harnessing the environment.

The wealth of data thus to be had is so limitless that we cannot afford to fail to develop the field to its utmost.

Present Patterns of Support and Their Insufficiencies
It can fairly be asked why any unfulfilled need exists at all in this field. The United States has led the world in the development and use of computers, and every major university now has a sizable computer complex that supports its research and educational endeavors. These facilities must serve and satisfy a wide variety of users. On the one hand, literally thousands of students are working problems and learning to use the machines. On the other, besides theoretical chemists, there are nuclear physicists analyzing results of accelerator studies, astrophysicists computing the properties of the sun, economists studying with mammoth computations the vagaries of the economic scale, English scholars making anthologies, linguists trying to learn how to translate automatically from one language to another.

Users of university computing centers therefore fall into two distinct classes that we may typify in brief as "large" and "small" users. Most users have small or relatively finite limits on the size of computations that can be made usefully. Theoretical chemistry, however, falls into another category, for it is evident that it can utilize effectively the whole of the very largest computer that can be provided, now and in the foreseeable future.

Each university administrator has a serious and urgent problem in facing the decisions of providing and financing computing facilities for his institution. It is well known that although computation is becoming cheaper in the absolute sense (each single multiplication or addition costs less money), the total cost of a computing establishment has risen now to the point that it exceeds the capabilities of individual institutions to provide facilities they now sorely need.

Up to the present the usual pattern has been that the university secures the largest possible computing system it can get within the limits of its finances in order that those investigators who really need the large systems can have them. Thereby, however, very frequently much of the system is being operated inefficiently by its small users. Time sharing furnishes a prospective mechanism for transcending at least some of this inefficient use of large systems, making it possible for heavy computations to occur as a background mode to smaller ones. It is unfortunate, however, that even this kind of use is strongly discouraged by inadequate and outmoded accounting practices which encourage and sometimes force the computing center manager to shut down his machine in order to maintain an artificially high hourly use charge.

On the other hand, to those investigators who are machine and time limited (as are theoretical chemists), being placed in the background mode is itself a strong limitation to the progress that can be made. The whole machine is needed. The whole machine in such a time-shared system is never available.

The practice whereby each institution has its own general-purpose computer also leaves much to be desired in the allocation of machine time among problems if the computer is busy almost all the time. Is a problem on the computation of the properties of a biologically active molecule more or less important than the study of the sun, than the construction of a Shakespearean anthology, or than the education of more students in the use of computers? Who, in a university, can make this decision in a rational way? Since we are reaching the limits of the university's financial capabilities, the decision must be made. Because of the financial stringency of university budgets, many of the real opportunities yet to come in theoretical chemistry will be lost to us through failure to provide the computing facilities needed, and even when they are in principle available, they may be lost because of auditing stupidity.

Finally, we should mention the inability of present facilities to provide a sufficient critical mass of experiences and capable theoretical and programming talent needed for the development of complicated software systems for theoretical chemistry. Graduate students and post-doctoral associates can and do provide an important resource in these system development projects, but they cannot provide the continuity of development required overtime. Professional appointments, with their wide-ranging responsibilities, do not provide sufficient time to perform this development either, and it therefore seems essential that the National Center have associated with it a permanent faculty of modest size with concentrated interests in planning and development of such systems. Only in this way can really effective use of the full capability of the available computer systems be made.

29

User Organizations
Report of Workshop 7

MARTIN GREENBERGER
The Johns Hopkins University
Faculty Discussion Leader

Contributors:

PAUL BARAN
Institute for the Future

HUGH CLINE
Russell Sage Foundation

WAYNE COWELL
Argonne National Laboratory

CRAIG DECKER
The Johns Hopkins University

JONATHAN J. KING
National Institutes of Health

ARNOLD LIEBERMAN
Rand Corporation

VERNON M. PINGS
Wayne State University

SALLY Y. SEDELOW
University of Kansas

ROBERT SEIDEL
Human Resources
Research Organization

WARREN SEIDER
University of Pennsylvania

JOHN R. SENIOR
National Board of
Medical Examiners

FRED TATE
Chemical Abstracts Service

Synopsis

In discussing how users can or should organize to facilitate the development of computer networks, the workshop examined the concept of a user group and established a few related definitions. It reviewed the factors that favor or impede the formation of user groups and explored how these factors might be affected by networking. After discussing policies and mechanisms that are needed to protect user interests as networks grow, the workshop considered steps that might help in the development of a national network and the impact this network might have on its users.

Costs as well as benefits arise from membership in user groups, and networking has decided effects on both. The costs that networking threatens to increase warrant special attention. In setting network policies, the user's interests are best served by measures that help keep the network flexible and self-organizing according to needs. Extensive feedback and critical evaluation of performance are essential to assist network development. External funding can be provided either to user or supplier. It might be best to direct most of it to suppliers in the early stages but increasingly more of it to users later on. Mechanisms are needed to ensure high-level documentation and maintenance of network resources. A device analogous to page charges is a possibility. An operating committee of user representatives for the network would help assure that network management does not shortchange their interests.

Definitions

A *user group* is a coalition of computer users who share resources—hardware, software, data, expertise—to achieve a common goal. In its most loosely organized form, a user group may resemble a marketplace for exchanging resources, such as statistical packages. There is little coordination or communication among users, and goals are common only to the extent that the exchanged resources have similar uses. At the other extreme is, for example, a team of meteorologists and other scientists all working toward a common goal, specialized according to the individual skills needed, and in frequent communication, who confer over a national computer network.

Most user groups fall somewhere in between these two extremes. Their purpose may reflect a number of common needs or desires.
- To share resources not available or easily maintained locally.
- To justify the preparation of program products through widespread dissemination.

- As producers of research-oriented programs and instructional materials, to achieve a "critical mass" of talent.
- To distribute program development, consultation, and other costs.
- Through standardized program documentation and languages, to enhance program products.
- To make known current research results, including new methods for computer-based instruction.
- Through the establishment of standards, to improve the reliability of shared data resources.
- To reduce the cost and time required to gain access to data, which can be critical when (for example) in the contesting of a new patent application, the patent is issued automatically if response is not made by a certain time.

Some examples of user groups, with the most formal and permanent groups presented first, are the following:

- Quantum Chemistry Program Exchange: permanent and formal; distributes user-supplied application programs.
- SHARE (IBM user group): permanent and formal; interface between IBM users and the manufacturer.
- National Board of Medical Examiners: permanent and formal; responsible for evaluating the knowledge and competence of medical students and physicians for certification.
- CACHE (Computer Aids for Chemical Engineering Education) Committee: permanent and formal; National Academy of Engineering committee to accelerate and coordinate the introduction of digital computation in chemical engineering education.
- CONDUIT Project: temporary and formal; cooperative effort by five regional networks to test the feasibility of transporting computer-based instructional materials.
- IMSL (International Mathematical and Statistical Library) Users: permanent and informal; informal interface between the program supplier (IMSL) and the users.
- RBS (Research for Better Schools): permanent and informal; nonprofit organization working with 300 schools nationally to develop and disseminate individualized instructional materials; the computer administers tests, diagnoses, and prescribes instructional plans for teacher implementation.
- Individual scientists and educators interested in collaborating in projects to cross-validate an instructional program for a particular target population; temporary and informal.

Not all sets of individuals who might be expected to form a coalition do so. This may be because of the implicit "costs" involved, because of natural inertia, or simply because they have not yet gotten around to it. A potential user group (PUG) is a set of individuals who could benefit from forming a coalition. In terms of PUGs and user groups, a *network* may be defined as a collection of potential or actual user groups, the resources they share or could share, and the communication channels that connect them to these resources. Subnets are the separate user groups within a network, together with the complex of software, data banks, and other computing resources and personal contacts to which they relate.

Members of user groups, through the roles they play in other organizations, can have direct effects on network activities in ways other than as members of user groups. They may

- Participate in university and institutional computer advisory committees and influence budget allocations for networks.
- Advise funding agencies by refereeing proposals.
- Participate in professional societies' policy making.
- Provide materials for publication.
- Make recommendations on product utility and policy to colleagues in their disciplines.

While drawing membership from traditional disciplines or professional society organizations, user groups may retain an independent status and have separate meetings. For example, relevant to the topic of computer uses in instruction, special interest groups such as SIGCUE of ACM and SIGCAI of American Educational Research Association draw on psychologists, computer scientists, educators, and engineers, all of whom have their own discipline and parent professional society.

Possible Advantages and Disadvantages of User Groups

In addition to the reasons already noted, user groups have been formed and people have joined them in the past in order

- To provide weight for the exertion of political pressure.
- To increase individual productivity through faster feedback to one's scholarly activities and greater opportunity for collaboration among colleagues of similar interests.
- To increase continuity under conditions of the mobility of faculty members moving from one institution to another, scientists transferring from one business firm to another, and graduate students leaving the institution from which they took their degrees.

- To gain credit within one's organization for participating extramurally in the user group.
- To obtain economy of scale.
- To provide a potential vehicle for training and funding.

Among the reasons why user groups have not been formed or people have not joined are the following:

- Fear of "premature closure," or of increased interaction's inhibiting a diversity of approaches to a given problem.
- Fear of "premature disclosure," or of a researcher's disclosing his strategies and results before he has had time to publish them and gain credit for his work.
- Fear of feeling pressured into using someone else's procedure rather than developing one's own.
- Uncertainty about the reliability of data made available to a group and about its hidden assumptions.
- Reluctance by a person who could make data available to the group to take the time for explanation and further definition, and competing demands on one's time in general.
- Fear of the government inferring illegal collusion.
- Lack of control over the actions of the group and need to compromise one's own goals and learn new conventions.
- Proprietary constraints.
- Dislike of travel and increased travel and communication requirements.
- Commitment and resource contributions required for participation.

Impact of Computer Networks on User Groups

The workshop reviewed the functions presently performed by user groups and considered whether the performance of these functions would be improved or worsened by the addition of networking.

Those functions seen to be clearly improved by computer networks included

- Allowing the user to get information, software, and computers not available within his home organization.
- Allowing access to common equipment and resources.
- Developing new resources.
- Increasing individual productivity, including improved capability for external collaboration and improved individual mobility in research, where one is able to retain access to data bases and tools developed by others.
- Reducing travel time and the need to travel.

- Providing better access to the programs and data of others.
- Overcoming some of the restrictions of local processor differences and improving confidence in one's data bases by providing checks on reasonableness.

The workshop saw as possible negative entries an increase in the danger of "premature disclosure" and a rise in the overall costs of interaction, although it was far from certain about these effects. It also questioned whether the prestige of interaction via a network would provide "Brownie points" equivalent to those provided by physical travel and whether electronic collaboration would be as enjoyable as physical attendance at meetings. The workshop felt that the forms of interaction would certainly be different and wondered if the opportunity for chance meeting of individuals with common interests might conceivably be reduced.

There were other factors the workshop thought likely to have a significant effect, but it was conceded they could go either way depending on the details of the situation. These included

- The pressure group function.
- Reaction time.
- Ratio of time spent working to time spent communicating.
- The user's feeling of effectiveness or powerlessness.

Policies Needed to Protect the Interests of the User

The workshop sought to identify some of the important policy issues in networking as they affect the user. The philosophy adopted was to examine incentives and policy guidelines that would allow users to organize themselves to the extent they want according to their interests and needs. The policies and mechanisms that appeared to be required for a user-driven system included

- Extensive feedback and evaluation mechanisms.
- Flexibility in system design.
- Focus on users in the development and composition of a system of many overlapping subnets.
- An understanding that users do not have to be organized in order to enter the network, but once in, may be affiliated with different subnets in a variety of ways under an open membership policy.

With respect to the funding of computer network development, the question arose of whether the user or the supplier, or both, should be supported. Users could be funded through vouchers used to supplement their own money. Suppliers could be supported through research grants that buttress or precede the financing they receive from network revenue. One interesting possibility is for funding at first to be directed primarily to suppliers (or potential suppliers) to promote their development of networks and to shift gradually to users to encourage development for their needs and purposes, and to stimulate healthy competition among suppliers. A possible source of this support, in addition to grants from government and private foundations, would be a tax applied to business flowing over the network, applied, once again, either to users or suppliers or both.

A sensible policy for computer network development would seem to be to direct initial efforts toward target areas where the following conditions hold:

- A need for resource sharing is recognized.
- Some computing resources and services exist.
- A private sector is available that could be brought in to perform those functions private firms do best (marketing, dissemination, and so on).

These conditions are the ones under which users are most likely to want and be willing to organize themselves.

There is need for a mechanism to ensure the continued documentation and maintenance of program and data resources on the network at a high level. One possibility is a device similar to page charges in the publication of professional articles. This might provide, as part of a research and development grant for a computer network resource, for funds to be used for its documentation and maintenance well into the future and well beyond the termination of the work required for its development—in effect, a fund for its "perpetual care."

Of great importance in the drafting of research contracts is provision for mechanisms of feedback and evaluation for gaining information on

- Who is using the computer network for what purposes.
- What new services and capabilities are desired.
- What are the problems in service.

This information would be useful for computer network management, and it would be essential to help guide future research on the computer network. Such research cannot be done well without a good understanding of user interests and needs.

An interesting question is whether there should be editorial review of material to be made available on the computer network. In the case of large-scale use of instructional material, it seems imperative. But for research purposes it could be overly restrictive and could lead to user groups that are too formal and overbearing. A better approach for research would be a system of information and communication that made users aware of new materials and who was using them, noted known errors, and provided a general market review.

Other policy areas that need much greater attention than was possible for the workshop include the following:
- The extent to which certification of materials is desirable.
- How scarce computing resources should be allocated.
- The volume and content of traffic on the computer network.
- Entrance requirements for new users and suppliers.

Perhaps the best way to secure a system that is controlled by the user and oriented to their needs and interests would be to have an operating committee made up primarily or exclusively of representatives of the various user subnets that make up the using population of the network. An advisory group made up of executives and officials who are not themselves users might also be desirable, but the user committee would be the primary body responsible for operating policy and for making certain that developments were "on course" from the user point of view.

Steps to Initiate User Participation
To implement user participation in computer network activities and to assess the impact of the network on user functions, the workshop suggested a series of steps to initiate network operation and then to expand or continue operation.

1. Identify potential user groups. Within the scientific research and educational community, users of data processing need to be identified who might profit from communication with other users to share data bases, programs, expertise, or machines. To identify these users and to provide network management with an appreciation of the scope of the potential user market and requirements, a variety of approaches are necessary:
- Canvassing professional and scientific societies, commercial groups, and trade associations, for potential computer network participants.
- Surveying universities and research institutes for recommendations on potential users who might benefit from computer network activity, including groups with specialized functions.
- Announcing and describing the computer network through the scientific press and inviting users to indicate their interest and requirements in forming new user groups.
- Circulating reports from existing regional networks, ARPANET, and

the Lister Hill National Center for Biomedical Communications network concerning present user groups.

2. Find out what these potential user groups are doing now and what they would do with a network. Define user activities in terms of the information-processing job they are doing and their requirements at present and under computer network operations.

3. Initially select users with a high degree of interest in, readiness for, and probability of benefit from computer network utilization. Study in detail the functional specifications of these users in their use of the computer network. Assess the costs and effectiveness of their present operations. Estimate the benefits and costs of computer network utilization in accomplishing their jobs.

4. Determine the initial set of user groups to participate in computer network implementation based on objectives of the network and judgments as to which user groups are most likely to utilize the network successfully. Analyze the costs, time, and personnel required to develop the initial network to an operational stage, including documentation and testing. Design and implement the computer network. Collect actual data concurrently on present user operations. Measure how well user functions are being accomplished with the existing system, and measure its costs for comparison with later operations using the computer network.

5. Upon initiation of computer network operation, assess the extent to which the user is able to accomplish his job better, faster, and cheaper or to do things not previously possible. Collect data during network operation, using measures previously employed and perhaps developing new measures, as appropriate, for future reassessments.

6. Observe and evaluate the consequences of computer network utilization, including changes in behavior, attitudes, and group structure for particular user groups. Perform an overall evaluation of all active user groups, and of the feasibility for computer network expansion, addition of users, and scheduling, based on data from the initial pilot program.

7. Recycle through steps 1–6. Have computer network management report the results and implement network operations for additional users as guided by the data obtained and the policies developed.

A careful program of monitoring, evaluation, and improvement based on what is learned about operation of the network and its effects will help keep development of the network headed along a course of ever better service to users.

30

THOMAS E. KURTZ
Dartmouth College

The NERComP Network

First, two definitions. I will use the term network to mean a connection between three or more institutions that are more or less independent and sovereign. This definition excludes single-supplier services, which exist in profusion, whereby a larger institution provides some form of computer service to the surrounding countryside (which may be quite large). It includes, in addition to NERComP (The New England Regional Computer Project), such organizations as MERIT in Michigan.

The second term is regional, which I will define as something less than national. In most cases the natural region is the state, by virtue of the state government. In New England the natural region may well be the six New England states.

I will now describe NERComP, which is the current manifestation of the desire among institutions in the New England states to improve the quality of computer services available to them.

NERComP, Inc., is a nonprofit corporation that delivers a variety of computer services provided by several (currently five) different university computer centers in New England to a number of users (currently at fifteen institutions). Many of the users are at small institutions that have no computing resources of their own. Others are at larger institutions that have a computing resource but cannot economically provide certain specific desired services.

The antecedent of NERComP was the New England Regional Computer Center, which was established in 1957 at MIT and which used resources granted by the IBM Corporation. Under this arrangement, any New England institution of higher education could use the MIT computer during the second shift. Many institutions took advantage of this opportunity, and for most of them it was their first exposure to computing. During the next ten years many of these institutions obtained their own computers but continued to use the resource at MIT for services that were specially tailored to that machine and that could not easily be transferred or transported to other computers.

In 1967 the ten-year agreement with IBM expired, and the New England computer users met with MIT officials to discuss ways in which computing might continue to be available to them under different funding and access arrangements. Since by this time a number of centers in the region had computer service to distribute, ways were sought to obtain funds that would allow smaller institutions to purchase there services on an open market. Accordingly, MIT prepared a grant proposal to the National Science Foundation on behalf of the New England college and university users. At about the same time the NSF embarked on its own plan for supporting cooperative computer projects. In a course of discussions between the NSF and MIT the original proposal was significantly modified to become the NERComP project.

The project's purpose was to study the pattern of use and the nature of demand for shared computer services. It sought to learn what kinds of service were needed and what kinds could be provided and shared over a network. At the same time it provided a small amount of funds for users, primarily from the smaller have-not institutions, to obtain computer service from the university computer centers. Accordingly, a simple telephone line network was established.

Originally a user would call NERComP and ask to be connected to some specific computer service. He would not necessarily have to know rules for gaining access to that computer. Later the human operator was not involved; users were connected to computers according to a prearranged schedule. In the beginning, heavy emphasis was placed on simple computing so that the basic service could be provided fairly cheaply and conveniently to small campuses in New England. This aspect of the project became less important as other, more economical sources of computing became available.

During the middle of the research effort for the NSF, NERComP became incorporated. By this time NERComP was acting primarily as a marketing agent for its suppliers' services. When the NSF grant was completed at the end of 1972, this marketing activity had allowed NERComP almost to achieve financial self-sufficiency, with gross revenues approaching $8,000 per month and with income from membership dues of about $20,000 annually.

From the NERComP experience to date, several important lessons have been learned. The first relates to the marketing of the simple computer resource to have-not institutions. NERComP here acts simply as an agent for one or more of the supplying centers. This activity has helped small institutions get started. The total cost to such institutions is small, inasmuch as very small quantities of service are involved, and long-range commitments are not needed. Such marketing is not viable in the long run, however, for two reasons. First, the market is self-limiting and diminishing, as alternate and cheaper sources of simple computing become available. Second, this marketing may be in direct competition with one or more of the suppliers, who may be carrying out vigorous marketing activities of their own.

On the other hand, NERComP's distribution of highly specialized services is increasing and will almost certainly continue to grow. The need for and variety of these special services seem to be increasing at a rate far greater than an individual center's ability to mount and maintain them. Thus, their distribution is likely in time to become NERComP's principal activity.

An example of such a special service is that provided by IMPRESS, a system for analyzing social science data that can be used by large numbers of students simultaneously. While quite reasonable imitations of IMPRESS have been devised elsewhere, the difficulty of offering the full menu of analysis techniques and access to a large collection of data bases precludes running it on any but large systems. But many faculty members at small schools wish to use IMPRESS in their teaching and research. Offering such a service over a network appears to be the solution.

Another example of a specialized service is a special computer system such as MULTICS at MIT, DTSS at Dartmouth, or CP/CMS at Brown. Each of these has special features that might be needed by faculty members at other institutions. If their requirements are less than would justify a full-time connection to the desired machine, then supplying such a service over a network appears desirable. Other activities, such as data-gathering projects carried out jointly by several institutions, appear to be practical over a network.

Another issue uncovered by the NERComP project relates to technical support. Whereas a much larger effort or one based in a university or research organization could count on a continuous source of technically qualified assistance, a small and independent project such as NERComP cannot easily command that kind of support. The solution appears to be to work toward better cooperation with the suppliers, most of whom have good technical staffs.

Perhaps the single most important suggestion for the improvement of the sharing of computer resources and service in New England is to return the brunt of the marketing efforts to the suppliers, most of whom already have a built-in customer base through their local networks. The local center would encourage its customers to use network services and in return would receive some sort of compensation. The point is that these local centers are, or should be, actively marketing their own services within their institutions and the surrounding area. If the compensation is right, it should be possible to achieve a balance between services provided locally and those obtained from other sources on the network. Further, the NERComP organization would be competing less and less with its own members and acting more and more as a coordinating agency.

Just as organizational structures should reflect the tasks being done, a network should reflect the true relationships between the nodes. In this case, the network of personnel supporting user services at the local supplying centers is the natural one. NERComP's role should therefore be to coordinate their activities, get them to talk to one another, and encourage them to develop new and better documentation and in other ways to improve the quality and variety of the services offered. NERComP itself should avoid, except in unusual circumstances, developing programs or documentation but should serve primarily to encourage their development and to provide some sort of quality control. NERComP would also provide, in its capacity as coordinator of the network, central billing and accounting for the services provided over the network.

A crucial factor in any project's success is that of critical mass. There must be sufficient use to justify the fixed expenses. If one's goals are to supply simple services, the number of users needed is apt to be very large. But if more sophisticated services are being offered, the critical components are the number and variety of these services. No one of them would normally generate enough interest to justify a network, but taken together a large number of them could provide the traffic to justify establishing a network and its continuing support.

In NERComP's case, the present type of traffic, based as it is on only a few specialized services, provides enough gross revenue (about $8,000) to support more than half its costs. The proper direction seems to be to encourage provision of a greater variety of services rather than to attempt to enlarge the customer base for the current services.

While it is not the purpose of this paper to discuss technical matters, there is one technical limitation that must be alleviated. The present switched-line network is preventing further network development. Besides limiting the number of possible interconnections (each interconnection requires manual action), data rates are fixed by the modems and multiplexing equipment. A switched-message network operating through a switching computer would alleviate these problems. Not only are all the $n(n-1)/2$ interconnections easy to make, but variable data rates can be easily accommodated. Such a system would also provide error detection and retransmission, an important factor in certain parts of the region where the quality of telephone lines is unpredictable. Other characteristics of a switched-message network are that it degrades gradually as traffic grows and that it is fairly easy to expand its capacity (by replacing voice-grade lines with ones having a higher capacity, for instance).

It should be made clear that an ARPA type of network with its distributed capacity and route-selecting features is *not* being suggested here. However, many of the features of the ARPANET have value in smaller networks; these include error detection, independence from the actual data rate of each terminal, much better measurement of traffic, and the ability to increase capacity without concerning the users.

An important point to be made is that expanded regional networks such as NERComP may well be the practical way to provide an eventual national intercomputer, interinstitutional network. Certainly the problems that will be encountered in a region will be a subset of those to be expected at the national level; the regional solutions will therefore be instructive. Working with a relatively small region may also be the sensible way to make sure that the needs of the small colleges are being considered. Finally, the time necessary to develop operating regional networks may be much less than that needed nationally. One important proviso: the regional networks must be coordinated well enough to ensure compatibility.

Based on its experience and the experiences of its supplying local networks, NERComP is developing a plan for an improved regional network along these lines.

1. Cooperation among the member institutions, especially the suppliers, will be stressed. NERComP will primarily play a coordinating role.

2. As much as possible, free marketplace ideas will be promoted; each user will be free, for instance, to shop around for the best statistical package for his purposes.

3. NERComP will coordinate each institution's user-service personnel, who will be the primary source for information and documentation about services.

4. Local network operators will market to their customers not only their own services but those available from the network. The local center will receive some form of compensation for all actual use of network facilities generated from within its local customer base.

5. Advisory boards will be established to coordinate the development of new services (to avoid unnecessary duplication) and to establish standards for documentation. Each local center would, however, retain

its right to develop whatever systems it wished, since the agreements reached in an advisory board would be suggestive only.

Although it is too early to say for sure, sufficient experience has been gained to suggest that the NERComP model is a viable one for the sharing of computer services and resources. Despite problems deriving from start-up, technical difficulties, and (lack of) economy of scale, its record of achieving near self-sufficiency in two short years of operation shows that there is a serious and felt need, one that is backed by funds even in these tight times, and that NERComP has gone a long way toward meeting it. Much more has to be done, but the network strongly suggests a useful direction to pursue.

31

Regional Computing Systems

Report of Workshop 8

JAMES L. MC KENNEY
Harvard Graduate School
of Business Administration
Faculty Discussion Leader

Contributors:

SAUL AMAREL
Livingston College of
Rutgers University

NIELS H. ANDERSON
University of Nevada

BAYLISS CUMMINGS
Flathead Valley
Community College

JAMES FARMER
California State
University and Colleges

PAUL GILLES
University of Kansas

ALEC GRIMISON
University of Puerto Rico

ZOHRAB A. KAPRIELIAN
University of Southern California

THOMAS KURTZ
Dartmouth College

LOUIS D. PADULO
Stanford University

CHARLES R. VAIL
Southern Methodist University

CHARLES WARDEN
Data Resources Inc.

DAVID WAX
New England Board of
Higher Education

The initial discussion of this workshop considered the role of existing regional networks. Defining them as geographic coalitions of users was regarded as too limiting, since most of the existing regional systems came about from a series of factors not necessarily dependent upon geography.

The discussion then considered successful regional networks and what was important in their development. It would seem that TUCC, NERComP, MERIT, and other successful examples exist as a result of four factors. First, an entrepreneurial leader coalesced opinion and developed an initial cooperative spirit among different institutions. Second, external forces in the form of budgetary constraints, obsolete technology, or the encouragement of cooperation by funding agencies acted on the normal institutional computer systems. The third factor was a political force, often in the form of a large organized user group, which lobbied for new or broader organizational forms to supply computing service. Finally, there existed a technological opportunity such as a larger computing device and a communication system that could make it available to all users if they cooperated. The development of a communication system was typically the only really new factor, as the other three ingredients already existed in one form or another.

Given these four factors, all successful networks evidenced a specific commitment to providing good service to a broad group of users, including documentation and reliable delivery of the service. A common aspect was some form of organization that began informally but gradually became more and more contractual in nature. The informal evolution seems to result in each user's being convinced that he had influence in the overall delivery system. A critical aspect of all networks is that funding and charging procedures are well defined and seem fair to all the participants, having been approved and in large part initiated by a participatory style of decision making. Finally, each of the present networks without exception began with some seed money to put the communication system in place and to give free time to the entrepreneurs to begin their efforts.

Developing New Networks
The discussion then moved to how to grow new networks. In the process it identified a class of system users defined as a potential user group (PUG). It is a set of users that is not organized at present but that represents those who will contribute to the growth of networks in higher education. In a way, any group exists because of something that isolates its members from everyone else. Thus, PUGs members might be iso-

lated by their discipline, by their demand for a special kind of service such as data base management or GPSS or graphic hardware, sociologically by (ethnic, economic, or educational) dependence on the budgetary control of some agency, such as a state or city, or by the government funding process through the National Institutes of Health, the National Science Foundation, or what have you. A more succinct definition of users and their characteristics is the following: (1) Those made cohesive by the control forces working on them. The forces may be political, financial through dependence on government grants, or an economic-sociological function. (2) Those organized by disciplines, since typically special service needs related to a specialist within a discipline. (3) Those without access to service for geographical reasons related to economies of scale.

Each of these PUGs has quite different goals and objectives and therefore requires an appropriate organizational capacity of its own. A PUG formed by some budgetary control force or funding agency typically comes into being with a formal structured information system and a set of time requirements for both reporting and relating activities to given institutions. It would appear that in the short run these goals and opportunities are to achieve the general level of computing service at the most economic rate for a wide variety of institutional goals. Formal organization develops by the direction of controlling bodies and appears to require information of alternative services. They do not need sponsors.

The PUGs that have grown up across institutional boundaries, such as that of theoretical chemists, have apparently appeared by spontaneous genesis as the inspiration of one or two dedicated individuals. It may be desirable for NSF to seek this type of growth in carefully selected areas, as there does not seem to be any well-organized, cohesive force to generate such organizations. This raises a number of uncertainties and interesting questions about leadership needs. Should each PUG define its own organizational structure according to its peculiar characteristics, or is there a general model? What mechanisms will permit the PUG to exercise maximum leverage as a consumer group over the suppliers? In general, disciplinary groups pose the most difficult problem in organizing to develop the software the discipline needs.

The needs of the third group, which is service-oriented in a general area, seem to have been serviced to date by the present regional networks and their evolution should be encouraged. An analysis of the needs versus service availability by geographical areas could probably identify future service requirements that could be met by regional networks.

Organization Issues

The growth of computer services from present regional network systems to a future national network will probably involve the development of a hierarchy of computing throughout the country. Some institutions might have no computing and depend completely on networks, while other universities might be nodes in a network and have special-purpose computers for local users with no network accessibility. The challenge is how, given a rich hierarchy of computing services, to organize the particular user groups effectively and attract them economically to the most efficient computer service.

The discussion on the developmental aspect of user groups in particular institutions pointed out that most PUGs start by acquiring adequate machine cycles to get some work done for a particular array of tasks. The next need often is for a set of higher-level languages that will allow particular individuals in an institution to move forward in their research and teaching. Finally, the users get to a stage where they require special-purpose software. It would appear at this time that networking per se may have the greatest demand from a reasonable-sized institution that has just developed a set of users with specialized software needs. Smaller PUGs that may not have the critical mass for machine cycles may start their computing on the network. As the users move from specialized systems into a network, they would evolve toward special-purpose languages and information services that would lock them into the system.

Selecting the Locale of Service

The workshop's discussion recognized many types of service but concluded that in general there are two extremes. One is general-purpose batch computing with straightforward operating systems, which has a simple economy-of-scale objective of obtaining the greatest amount of computing for the dollar. At the other extreme are special-purpose language systems encompassing a range of software and data maintenance costs that far exceed the hardware considerations. If service is available on a network, the economic decision of the particular computer user is: Should he initiate service on the local resource, or should he combine his demands with those of others to provide a specialized center? This decision conceivably could be made on the basis of balancing marginal productivity versus marginal costs relative to computing power and the service desired. However, as the dimensions of service become more complex and move away from the straight economics of technology and communication, such a trade-off becomes difficult to calculate, and the decision maker needs the help of an improved theory. In broad outline, such a theory would argue that the two independent measures are the level of use versus the complexity of the computer service desired. It would suggest that the data could be arranged in a form that would trade off these two metrics and would provide a scheme to decide which computing should be done locally and which should be done on a network.

This analysis reflects the concept that a wide variety of types of computer service exists and that it is not necessary that all of them be carried out equally well on any particular physical computer. Thus, a theory that purports to support optimal decisions must reduce all types of service to a set of basic economic parameters that can be optimized in relation to machine cycles and fixed and capital expenses. The conclusion of this discussion is that now is the time to begin collecting and analyzing a range of computer services in a fashion to provide guidance to local decision makers on what they should provide locally versus what they should obtain by networks.

An Evolutionary Approach to Networking
This evolution of networks suggests a hierarchical organization for the management of networks throughout the nation. The alignment would include local university or college organizations that have some relationship to a regional or disciplinary network. An alternative organization would be a nationwide system of regional networks and a set of nationwide discipline-oriented networks. This condition would pose to the local computer center's manager the issue of what service he should provide locally versus what services he should contract for on either the regional or the disciplinary network. The regional system would have to decide what particular specialties it should encourage within its region to service as a part of either a discipline or regional special services. These specialties would generally be those demanded throughout the nation for which a given region would have the most skill, technology, software, data, or interests to provide. The balancing of inflows and outflows of service within a given region and planning for the future are most difficult problems. This concept of an expanded regional network implies that most regions will have one or more specialties.

A national network might develop along particular information service lines. For example, it is conceivable that a national library network might develop by linking together a series of regional systems like that of the Ohio college libraries. At present there are two or three such

regional library networks. Linking them might produce further economies of scale and allow increased specialization in the development
of software and services. This service might expand to other areas such
as administration and the particular research and teaching experience
in faculties.

A discussion of the growth of PUGs led to an evolutionary strategy for
the foundation that seemed to follow the present development of regional centers. This strategy could be summarized as follows:

1. Link a few existing regional networks into a national system for opportunities to balance machine loads, to obtain backup support, to allow
a broad range of users to utilize special-purpose software at one of
the regions, and to acquire experience with networking between user
groups.

2. Encourage each node on the network to become specialized as well
as to offer general service to its region. For example, one regional network might develop a strong library service, another a rich array of
statistical systems, and a third a large list-processing capacity. Duplication
of special-purpose activities would, however, be discouraged.

3. Encourage new natural regional networks for general computer
services among institutions with common goals and geographical proximity. These networks would be formed to achieve economies of scale in
general-purpose computing but would keep an eye open for future
specialization.

4. Identify and recruit potential user groups that do not fit into a
natural network to join the national system for general as well as specialized computer service.

5. Create a national advisory board of regional centers, funding agencies,
potential users, and experienced users to appraise and guide the development of the national network. A key activity of this board would
be to identify and plan services for potential user groups.

6. Launch a nationwide user identification project aimed at defining
the specialized services a potential user group could afford to support and
identifying the special-purpose nodes a truly service-oriented national net
should provide.

Unusual User Needs

The user groups identified in this discussion as having a high potential
included the broad range of users identified in workshops 1–4 (see
Chapters 12, 14, 16, 18). The workshop felt that the stage of develop-

ment and the financial support available to the particular user groups
were key elements. In a university where computing is relatively advanced, there is a natural growth of computing and sufficient stimulation
and support for new users. For example, Princeton's annual growth is
about 20 percent and Rutgers's is about 30 percent. But in many four-
year colleges it is highly unlikely that computer demand will grow even
if computer resources are available unless there is a deliberate policy
of promotion and stimulation. These differences and stages of development have important implications for developing participative organizations. That is, high-growth users could dominate the planning and
provide a too sophisticated and therefore unattractive service that would
not serve to involve a broad set of users. Obviously, there are financial
issues in this growth and support of users. An important NSF role
would be to bring in users in nongrowing fields in coordination with
those in growing areas to see if demand could be stimulated and computer systems used more efficiently.

In addition to the needs of higher education, an extracurricular void
exists in most communities that could be filled, at least partially, by
local PUGs. Thus, a network might provide a community service for
agencies such as urban planning, community development, and schools
without walls. An established system could provide such service with few
incremental costs. Local governmental agencies frequently need research
or data-processing assistance on research projects, analytical activities,
or even straight calculating. Such cooperation might be applicable only
in certain situations where state facilities are inadequate. In general,
computer facilities on a network should be accessible to communities
where relevant.

A special condition of regional networks is apparent in Minnesota,
Nevada, New Jersey, and perhaps California. This involves the efforts of
state governments to consolidate state computing, including that of
higher education. These statewide networks may provide an additional
building block for a future national network but may also represent
difficult barriers to true cooperation. The political entity is very sensitive
to control issues, particularly budgetary ones. Different state institutions
encounter strains over sharing a facility, and interstate cooperation
could be difficult. But these agencies or state systems might well be good
customers for particular discipline-oriented special-purpose computing
if the state systems would include the discipline services in their planning. How to trade into these markets is a political as well as economic
problem.

Conclusion

A conclusion of this approach might be that the foundation should expend funds for purely distribution functions and not the support of hardware in the next decade and that entrepreneurs might be encouraged to submit proposals that would include not only a brokerage function but the discovery and formation of user groups from existing poorly serviced communities. This function would include identification of individuals geographically dispersed within one discipline, institutions not having adequate computer support, and isolated campuses that would feel comfortable in committing resources to the establishment of cooperative ventures. It would appear that this strategy involves a two-pronged approach, one proselytizing and organizing new users to make them amenable to cooperation and organization with a network, the other the organization and further specialization of existing computer network systems.

IV

Operations and Funding

Part IV contains the papers and workshop discussions of the final seminar on operations and funding. It builds on the ideas developed in the previous seminars but also rethinks some of the issues already considered in interesting new ways. Participants consider such questions as

• What research is needed to resolve critical questions in network operations as they have evolved so far? How might such research be organized and conducted? What criteria should be used to evaluate the results of the research?

• What design and operating principles should be used in the development of a trial network, and what major issues and critical problems need to be addressed in planning for the network?

There are five papers. Richard Bolt discusses the challenge of managing computer networks and draws a distinction between three different types of networks that serve complementary purposes (Chapter 32). Joe Wyatt presents the plan for external purchase of computational services currently being tried at Harvard University as an alternative to running a large central computing operation (Chapter 33). Gerard Weeg describes a number of regional star networks in different

parts of the country and draws some lessons from their experiences (Chapter 34). Davis McCarn presents the history and current status of the Medline Network for online search of the medical literature (Chapter 35). Randall Whaley discusses the promotional and economic problems that he has had to face in trying to bring a group of colleges and universities together into a computer-sharing organization (Chapter 36).

The four workshops are numbered 9 to 12. Workshop 9 considers alternative network designs from the hardware point of view, including computer equipment configurations, telecommunication approaches, switching systems, and basic network layout (Chapter 37). Special attention is focused on the capital costs of alternative approaches and their implications for operating methods, costs, and the ability of the network to serve important needs. Also considered are questions of telecommunications policy that conceivably could have a crucial impact on the economics of alternative network approaches.

Workshop 10 deals with the characteristics of operating systems and user protocols (Chapter 38). Considered are questions of design criteria, degree of coordination, standards, and system maintenance. Also discussed are the operating procedures of host computing centers, the contractual relationships between hosts and network, standards of availability and reliability, and methods for maintaining privacy and for verifying legitimacy of access.

Workshop 11 deals with the development of applications programs, data bases, and other services on the network (Chapter 39). Particular attention is paid to the processes by which new products can be identified, implemented, and marketed to prospective users. The nature of the service package that can be supplied is discussed, with emphasis on the need for good availability, a high level of documentation, and generally strong user support services. The workshop considers how to identify the need for services, how to deliver them, how to maintain application programs, and how to operate consulting and training programs.

Workshop 12 examines the economics of the networking business (Chapter 40). What are the advantages and disadvantages of alternate financial organizations and methods of recovering costs from users? How can economic incentives be used to help balance supply and demand in networking, for both the short and the long run? What are the economic merits likely to be and what is the overall social value of networking in economic terms? Workshop 12 considers the kinds of financial leverage that will be needed during the organization and start-up phases of network operation. It discusses the kinds of capital requirements that can be foreseen and explores where this capital is likely to be found.

32

RICHARD H. BOLT*
Bolt Beranek and Newman Inc.

The Challenge of
Managing Computer Networks

I am not a computer specialist. I have not done research in computer science. I have not developed a computer system. I have not even invented a computer network.

I have, however, been involved in the business of computers in certain ways. My role has related mainly to corporate policy of Bolt Beranek and Newman, which started working with computers in the late 1950s. We first used computers to help our psychologists study human interactions with acoustic signals. Today our research in acoustics and noise accounts for about one-third of our business, and another third involves information science and technology, including computer science and systems development. These fractions serve to show that my computer involvement has been not only peripheral but also part-time.

We started moving toward networks in 1962 when we developed a small time-sharing system for use in research on artificial intelligence. In the mid-1960s we had a sixty-four-terminal system distributed between our Cambridge center and a Boston hospital to explore new uses of computers in handling medical information. For several years we operated a commercial time-sharing network along the East Coast. In 1969 we started working on the ARPANET project.

Against this background I have come to see the computer network business as one that offers unusual challenges to management. Especially I have seen the importance of being very clear about the basic purpose or function that a particular network is supposed to serve. So here I shall look more at what a network is for than at what it is made of, realizing of course that function and structure are intimately related.

Because of this orientation, I need to do some defining. I am going to divide the universe of networks into three functional categories: task-centered networks, signal transport networks, and communication facilitation networks. In physical reality the three kinds can overlap and intertwine; they can form subsystems of a single network. But in concept the three kinds of networks differ distinctly in their basic reason for existing.

A task-centered network, as I define it, is any set of computer facilities put together to help a particular community of users carry out their task

* I am indebted to Stuart L. Mathison for assisting me in preparing this paper.

of common interest. Although I might have chosen to use the more familiar name, application-oriented network, I call it task-centered to underscore my purely functional categorization. Some examples of tasks in the sense I mean are to do research in quantum chemistry, to teach electronics, to collect and disseminate biomedical information, to order airline tickets, to deliver health care, and to study linguistics.

A prime challenge of managing is to face up to the question, how do you measure success? For a task-centered network the basic measuring instrument, it seems to me, is the perception of the users themselves. Only a user knows firsthand how it feels to use the network for its intended purpose. Such a network is successful to the extent that the users perceive it as really helping them to perform their task effectively.

Using this measuring instrument, human perception, of course involves complications. Finding out how people feel about something takes astute observation and skillful use of attitude surveys. Because there are many users, a suitable statistical measure or consensus must be found. Again, even if the performance of a network is the "best" that the state of the art can produce, the performance might not come up to the expectations that the users held in advance. Nonetheless, the user's perception is the last word. Even if the reliability, response time, and other objective characteristics are "perfect," people will not go on using a task-centered network unless they come to feel that it helps them perform their tasks better.

This kind of network can stand alone as a collection of interconnected equipment serving several users, all located in one building. More relevant here is a set of task-oriented equipment and users distributed geographically at arbitrary distances. Then the second function I have defined comes into play: the signals must be transported from place to place. The third function, that of facilitating the communications, may or may not be added, so I shall discuss it last.

A signal transport network, in my semantic, is one that carries signals from one place to another at some distance. I arbitrarily exclude the wires that run around inside a single computer center. Examples I have in mind are dial-up telephone or Telex lines connecting a time-sharing bureau to its external users and leased telephone or telegraph lines interconnecting computers and terminals located at geographically separated sites. The network may utilize active elements such as repeaters and concentrators, but under my definition it provides only a passive function that amounts to signal-in–signal-out.

The effectiveness of a signal transport network is measured directly by how well it does the transporting. Relevant measures include the information rate, which may be expressed in kilobits per second; the reliability, which may be expressed reciprocally in terms of errors, distortion, or downtime; and the cost, expressed in terms such as cost per message, per bit, or per hour.

For assessing a signal transport network, the relevance of the users' perceptions is only secondary at best. If you can "feel" the network, it it is not really doing its job well. Ideally, a user should not be able to perceive whether the computer is right next to his terminal or a thousand miles away. What does count is the objective effectiveness and its relation to cost. The rational choice goes to the alternative that provides the required information rate and reliability at lowest cost.

The third functional category, the communication facilitation network, is one that enhances the process of communication as by making it more reliable, more economic, or more versatile. It lies between a task-centered network and a signal transport network and may overlap them physically. In the language of telecommunications policy, this kind of network is coming to be called a value-added network, but my name for it makes its function more explicit.

We can see what such a network can do by looking at a specific example. I am familiar with the ARPANET because my company produces its interface message processors (IMPs) and terminal interface processors (TIPs), operates its control center, and participates in further research and development for it. Actually, the ARPA computer network system encompasses all three functions that I have defined, but I shall focus on its facilitation function.

On the map of the ARPANET (Figure 32.1), each datum point represents a node in the network. The solid black points represent IMPs and the open circles, TIPs. Both processors perform message-switching or, more correctly, packet-switching operations, which route traffic through the network. Each IMP is tied into one of more user sites, each of which contains one or more "host" data-processing computers. Each TIP is tied directly to terminal devices through which users obtain access to host computers located elsewhere on the network. TIPs could serve local computers also.

The lines drawn between points on the map represent leased lines with a rated capacity of fifty kilobits per second (except for one line of higher capacity). These lines collectively make up a signal transport network

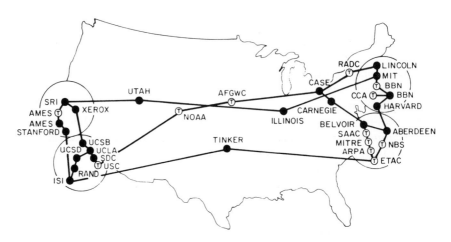

Figure 32.1. ARPA network geographic map, November 1972.

as I have defined it. This subsystem network can be extended by adding more lines, or using lines of greater bandwidth, or using satellite links, one of which has been added recently (and is not shown on the map).

The communication facilitation network, as a subsystem, starts physically at these nodes in the form of the IMPs and TIPs. These devices handle the switching and routing of packets each containing up to about a thousand bits and each carrying a destination address. When a packet arrives at a node, the IMP there stores a copy, forwards the packet to an adjacent IMP, and rubs out its stored copy only if it receives a prompt confirmation that the packet reached the next IMP safely. If a packet fails to get through or takes too long because that channel is over-crowded, then the packet is forwarded along an alternate transport line to another IMP. The several packets that make up a continuous message might take quite different routes to reach their destination. Once the packets are there, the IMP reassembles them and delivers the message to the intended receiver, a host computer. The TIPs do all the same things and in addition can communicate with a terminal device directly without going through a host computer.

Actually an IMP does not connect directly to a host computer. In between is a piece of hardware, a host interface, specially designed to match the IMP to the particular computer system involved. To provide further compatibility, the host computer itself contains a special piece of software called a network control program. These pieces of hardware and software serve functionally as part of the communication facilitation network, yet they reside in or alongside the host's local computer system that, in turn, may be a part of one or more task-centered networks. This arrangement is what I had in mind when I said the functional types of networks can overlap.

This overlapping poses an unusual management challenge. Who should control the parts that are common to the different networks? Any reasonable answer probably would include the stipulation that the host interface and the network control program should operate in accordance with standards and specifications that are controlled, or at least agreed to, by the management of the facilitation network. Facilitating communication among many computers, especially if they are of different kinds, requires some minimum degree of uniformity in the interfacing arrangements.

Returning to the map, we see that thirty-three sites are now tied into the ARPANET. IMPs serve twenty-one of the nodes and TIPs serve the other twelve. The first eleven nodes went into service two years ago, and since then other nodes have come on the network at an average rate of about one a month. The level of utilization also has been increasing quite steadily. A year and a half ago the traffic among nodes ran at about five kilo-packets per node per day; now the figure is around fifty. One packet can hold about the number of characters that fill one line of teletypewriter printout, although in practice some packets carry fewer characters than this, depending on the message being sent. The rate of fifty kilopackets per node per day is still far below the maximum traffic load that this network can handle.

Now that I have explained all three functional categories of networks, we are ready to see more explicitly how they can be put together to form a system. The diagram shows the essential relationships in simplified form (Figure 32.2). The inner circle represents the signal transport network, here made up of leased lines and related communications equipment. The adjacent ring represents the communication facilitation network and the outer ring represents the task-centered network. The user community works with the terminals (Ts), connected either to a host computer (HC) or to a TIP.

The network control program (NCP) resides in the host computer but serves functionally as part of the communication facilitation network. In some cases the host interface (HI) may be shared also, but the IMPs and TIPs are physical elements of the communication facilitation network alone.

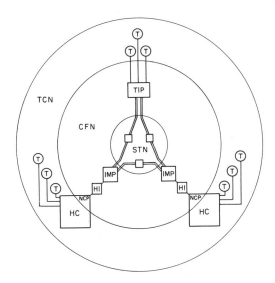

Figure 32.2. Simplified model of a computer network system.

The center circle contains the signal transport network (STN), including leased lines and communications equipment.

The inner and outer rings, respectively, contain the communication facilitation network (CFN) and the task-centered network (TCN), which physically share the network control program (NCP).

Interface message processors (IMP) serve the host computers (HC), to which user terminals (T) are connected. The terminal interface processor (TIP) serves terminals directly and does not use a host interface (HI).

A communication facilitation network can increase the reliability of communication by providing one or more alternate routes around a transport line that has become inoperative. Because the ARPA network is still under development and therefore is not yet being called upon to achieve its maximum possible reliability, its potential performance is not yet seen. Over a recent six-month period the downtime of the lines themselves averaged about 1.8 percent, and that of the IMPs and TIPs, including both their hardware and their software, just over 1 percent. During the last month for which I have data more than half the IMPs and TIPs operated with downtimes between zero and 0.4 percent. When this level of reliability is reached or exceeded by all the node equipment, the overall reliability of the network will become significantly better than that of the transport lines themselves.

Increasing the economy of operation is another way in which the ARPA network facilitates communication. The packets are switched and routed independently in such a way as to use the available signal transport network lines as efficiently as possible. Except for dead-end spur lines, essentially all parts of the network are available for use by every packet. The result is efficiency approaching theoretical limit.

This type of facilitation offers an unusual attraction. Because the distance between sender and receiver does not determine how far a packet may travel and because the cost of the lines themselves is only a fraction of the investment, the transport service could be charged for at a flat rate per packet independent of distance. Then a user community in developing its own task-centered network would not need to exclude a particular member just because he was located far away; the distance alone would not increase the communication cost.

Economy can also result from the sharing of computer resources among different centers connected to the ARPANET. Resources thus shareable include data bases, program packages, and computer power. This attribute relates to the third and perhaps most interesting way in which such a network can facilitate communication—by making it more versatile.

The ARPANET started out as a system serving a particular community of users, the persons working in computer science. Today these persons, though dispersed across the country, can work together in real time through the exchange of information and processing capabilities germane to their research. Their facilities collectively make up what I am calling a task-centered network. But the ARPANET is also serving other communities with other task interests, such as weather research. Thus, a communication facilitation network can simultaneously serve the interests of many different communities using computers of virtually any make or size and even including members who have only a terminal and no computer at all.

Suppose that a user community X has members at three particular sites. Suppose that a different user community Y has members at the same three sites. Because of the versatility of interfacing equipment such as IMPs and TIPs, a single installation of the equipment at a given site can serve both X and Y at the same time, without interference between them. So only three such installations are required instead of the six that would be needed if the two communities were to work through independent networks. A communication facilitation network is a common denominator. The number of interfaces is simply equal to the number of sites connected and not the number of sites times the number of task-centered networks that occupy those sites.

Because of this versatility, an existing communication facilitation network can be used as a test-bed for developing a new task-centered network. Two options are available. A research community can acquire its own computer equipment and use the communication facilitation network only for communication among its sites. Or the research community, through mutual agreement, can share in the use of computer equipment already employed by other task-oriented communities. In either case, each user community can itself decide who will receive its task-oriented messages and information and therefore can operate independently if it wishes to do so.

Now that I have explained what I mean by each functional category of networks and illustrated some of their features, we can look more explicitly at the challenges met in managing them. My central thesis is this: the three kinds of networks pose significantly different kinds of management problems.

To emphasize the differences, I shall consider each kind of network as if it were managed separately by an organization that involves itself with that one kind only. In real life, of course, a single organization can undertake to manage any two or all three kinds, in which case the problems met will be a combination of the problems that I will discuss separately. The problems themselves usually stem from technological, social, or economic factors, so these will serve as focal points for my comments.

Technology for signal transport networks has been available for a century in the form of telegraph and telephone lines. Associated communication equipment has reached a high level of development. The past two decades have seen the emergence of microwave links, both ground-based and satellite-based. Except for exotic refinements, the technology that these networks are likely to use for some time to come is in hand now or is predictable.

The notable challenge posed by the further development of these networks lies not so much in their technology as in their financing. Signal transport networks are highly capital intensive. They involve large-scale physical facilities that have a very long lifetime and are slow to change. How can such networks be financed at a magnitude and on a time scale that will enable them to reach an economy of scale adequate to provide a reasonable return on the investment? How can they be constructed to produce the compatibility and wide availability of service required to achieve economic scale? These questions are challenging indeed.

In contrast, much of the technology used in task-centered networks is changing rapidly and not very predictably. Cases in point are mini-computers, memory devices whose costs are dropping drastically, and terminals specially designed to serve particular task communities. Further, applications software can change on short notice and tend to evolve more or less continuously. So the management challenges of developing and operating task-centered networks lie more in the problems of coping with constantly changing requirements. The management must keep astute surveillance over the advancing science and technology. The resources in use at each point in time must be shared as widely as possible to hedge against obsolescence before the investment can be amortized.

Communication facilitation networks lie somewhat in between. Insofar as they take on characteristics of general-purpose common carriers, they must be financed with a magnitude and phasing of investment commensurate with economy of scale. At the same time, these networks must be managed in such a way as to keep them mutually compatible with the changing technology used by the task-centered groups whose communications the networks are endeavoring to facilitate.

Social factors introduce another set of challenges. New technology generally changes the way people do things and the way they work with one another. In turn, the changes experienced by the people, as they attempt to fulfill their needs in a different way, reflect on the technology and press it to undergo adaptive change. Technology is applied and accepted only if the market pulls it. Although cajolery and Madison Avenue can inject a dose of desire into the market, only the people themselves finally decide that they want the new technology, that they can adapt to it, and that they, in sufficient numbers, will pay the price that pulls the string.

Computer networks are here today, and they are not likely to go away tomorrow. Already they are changing the way we do some things, and they are likely to induce further changes we do not yet foresee. Let us look at some of the social challenges.

Perhaps the least affected within my three-part model of networks will be the purveyors of signal transport. They see the rapidly increasing demand for data communications and have projected, with what I believe to be justified confidence, an enormously larger demand in the decades ahead. They will have to expand their investments, capital facilities, and operations. But so long as the service they sell is simply more and better digital data transport per dollar, the managers of signal transport networks need have little concern for the details of social change taking place in the communities of task-centered users.

Users occupy the opposite pole; they are where the action is. Of great concern to them will be their task-centered network, its design, its technology, its management. In what respects might the users have to change their ways and attitudes? One is that they will probably have to give up some autonomy in exchange for wider opportunity. Today the persons using computer facilities exclusively within their own center may control their own practices and procedures. Tomorrow, when they share facilities with other centers through a network, these same persons may have to share control with the entire task-centered community, at least to the extent that compatible operation is needed to make the wider range of facilities usable by all members.

Some users may have to change their protocols and improve their documentation to support and expand the remote use of the computer-based, task-oriented services they are offering to the rest of the community. Again, some users may have to change their work habits to realize the potential value offered by the expanded capabilities. For example, the users will have to learn how to work together in real time, with various colleagues and computers located a thousand miles away. At some institutions the users may have to give up a cherished computer, or even an entire computer center, because they now can obtain better or less expensive service through the network. Such loss, in view of the attitudes it could evoke, will challenge users and managers alike.

The manager of a communication facilitation network will feel the effects of sociological change mainly through its impact on the market for his facilitation services. The manager will need to acquire a clear understanding of the interests and needs of the task-oriented communities. In close communication with those communities, he must join in evolving mutual agreement about the interfacing aspects of the task-centered networks and host computers, including protocols, standards, and specialized services desired. The network that he manages must be made compatible with a wide variety of computers, terminals, and applications software. And the network must be adaptable to the changing needs and habits of the task-oriented users while they, in turn, adapt their behavior to the changing technology that the network offers.

The manager's first problem is to understand the business objective of the particular kind of network he is managing. For his benefit, the following summary for each functional category will start with a statement of objective.

Task-centered Network

Objective: To provide a single task-oriented community with advanced, symbiotically adaptable computer capabilities to help its members perform their own task in their own way.

The manager will need to look on the network as an integral part of a task-performing system that includes people and their social interactions, intellectual goals, institutional affiliations, and procedures, as well as all the physical elements of the network. The manager will require at least an interpretive understanding of the technological and social factors involved as well as a problem-solving understanding of the economic factors. The market for each such network will be relatively small and specialized within the wider markets for education and research. (In this part of the summary I am ignoring networks for commercial tasks such as ordering tickets.) The business will be supported largely as a public good financed by grants, subsidies, and institutional funds.

Communication Facilitation Network

Objective: For each of several task-centered networks, to increase the economy and reliability of communications among its network nodes and to offer its users increased versatility in developing, using, and sharing their collective, geographically dispersed resources.

The manager will need to look upon the network as a general facility that is compatible with a wide variety of user communities, computers, terminals, and applications software. Management challenges will be met in adapting the service to a changing market of users and uses and in accommodating new customers with minimal imposition on existing customers and minimal new investment. To ensure long-term availability of service to the task-oriented communities, the business will probably require private financing of moderately large magnitude and time scale, although the continuing development of advanced technology may qualify for public support. Achieving long-term stability and economy of scale may call for a broadening of the market to include commercial customers as well as task-oriented communities in education and research.

Signal Transport Network

Objective: For a communication facilitation network as well as for many other customers, to transport digital signals over arbitrary distances, at low cost, with minimal error, delay, and interruption.

The manager will need to look upon the network as a major investment in proved technology to supply services for a large-scale, diversified market that promises long-term demand. The service is usually marketed as a private good. The business generally takes the form, however, of a regulated, noncompetitive public utility, although some change in this traditional pattern is being brought about, for example, in connection with the use of satellites. As time goes on, the manager will face the challenge of prolonging the useful life of old equipment while gradually introducing new technology. Raising and amortizing large amounts of capital and structuring the revenues to yield return on the investment as well as to cover the costs of development and operation will pose notable challenges to management.

This summary only touches some of the highlights. If people look at computer networks from the functional point of view; if they measure success in terms appropriate to the function; if they notice how technological, social, and financial factors impinge differently on networks serving different functions; if, for each of these factors, they think about relative magnitudes, time scales, and rates of change; I believe they will uncover an array of managerial responsibilities and challenges going well beyond the few specific ones mentioned here.

33

Harvard University

The Harvard Plan

My remarks come from the point of view of a person who has been at
Harvard for four months. As a result, some of the information I have
is based on historical record rather than personal experience. One of
these lesser-known historical records concerns an incident that allegedly
occurred during the summer of 1971 when Derek Bok, Harvard's president, walked into his office one morning to find an aged and tattered slip
of paper on which were written these words:

there is always
a comforting thought
in time of trouble when
it is not our trouble
       archy[1]

At that time several alternatives were being considered for the future
of computing services at Harvard. One involved the divestiture of the
major on-campus resources, and this, in fact, was the alternative ultimately chosen. We may never really know how much influence archy
had on Mr. Bok, but I trust you now have an expanded insight into the
secrets of the presidential decision-making process.

I will attempt to report on some things happening at Harvard that are
relevant for these discussions and to generalize them for the future.

First of all, Howard Aiken, as a graduate student and later as a faculty
member, developed one of the early digital computers—the Mark I. It
was followed by a sequence of similarly developed computer systems end-
ing with the Mark IV. These computers were very mission-oriented for
a cloistered group of people to do research in what amounted to computer science at that time. Other centers of interest developed through
the years and satisfied their computing requirements as best they could
more or less independently. In 1962, the Harvard Computing Center was
established, operating on a fee-for-service basis using "100 percent dollars." The center began using IBM 7000 series technology, later moved
to IBM 360 technology with an IBM 360/50, and finally acquired an
IBM 360 Model 65 for batch computing service. An XDS 940 and later
an XDS Sigma 7 were also acquired for time-sharing usage.

During the five-year period beginning in 1967, a $1.6 million accounting
deficit was accumulated. A year-by-year record of that deficit is interesting

Table 33.1. Harvard Computer Operations (000 omitted)

	1967–1968	1968–1969	1969–1970	1970–1971	1971–1972	Total
Income from federal grants and contracts	1,028.5	669.0	721.6	490.3	340.3	3,249.7
Income from institutional sources	603.3	842.6	1,003.1	1,023.0	851.8	4,323.8
Total income	1,631.8	1,511.6	1,724.7	1,513.3	1,192.1	7,573.5
Deficit	161.4	433.6	573.4	374.0	123.7	1,666.1

(Table 33.1). Notice that the deficit follows an almost normal distribution. You are free to decide the implications of that. Notice also that in the first year the income for federal grants and contracts was $1,028,000. Then it decreased significantly and virtually continuously during the following five years. For the same period, however, the institutional sources, primarily from instructional budgets increased steadily until 1971–1972. One could then theorize that if, for example, federal contracts and grants had sustained the income level of 1968, there would have been no deficit. Unfortunately, this level of analysis has limited value at this point. But if all other universities put their computer services on a full cost-recovery accounting basis, many could probably tell a similar story.

During the same 1967–1972 period several serious studies of all aspects of the role and scope of computing activities at Harvard resulted in a number of changes in the philosophy and organization of computing activities, including

1. The establishment of an Office for Information Technology to deal with the planning and management of computing and related technological resources.
2. The establishment of a Center for Research in Computing Technology to provide for academic instruction and research in computer science.
3. The divestiture of the major on-campus computer facility (IBM 360/65 and XDS Sigma 7).
4. Multiple arrangements for computing services on an annual contractual basis.

I might describe some of these arrangements. One of the first involved Harvard and MIT. They set up an equal partnership utilizing a 370/155 that began in November 1971 for batch computer service. Before that, in 1970, a three-year agreement was reached with First Data Corporation of Cambridge for the interactive services of a DEC system 10. At the end of a year of the partnership it was determined that Harvard had used approximately 35 percent and MIT 65 percent of the shared resource. As a result, for fiscal year 1972 the arrangement was changed so that Harvard became a major customer of MIT rather than a partner. In September 1972 MIT upgraded from an IBM 370/155 to an IBM 370/165. There is a record of the evaluation that led to the divestiture decision in a paper by John Austin of Harvard.[2] It was in this paper that a number of alternatives available to Harvard at the time are enumerated. Harvard subsequently chose the minimum risk alternative, with the help of archy, of course.

If we look at the situation now, we can support the statement that Harvard is quite committed to the philosophy of making major usage of off-campus computing facilities, including networks, to satisfy a majority of its computing needs. These resources include the IBM 370/165 at MIT, which is directly connected to terminals at Harvard including an IBM 370/145, which serves both as a remote batch terminal and as a stand-alone processor; a DEC System 10 at First Data Corporation with twenty-five dial-up ports; a new DEC System 10 at the Harvard Business School with twenty-four dial-up ports; and a DEC System 10 at the Harvard Center for Research in Computing Technology, which is used exclusively as a research machine and which is connected to the ARPANET. There is also an IBM 1130 in the Chemistry Department that is connected via telephone line to the IBM 360 Model 91 at Columbia. There is some modest use of NERComP, the New England Regional Computing Network.

I can say unconditionally that these computer services are working well. They are working at least as well as the facilities that preceded them and are significantly better in terms of overall fiscal health. The unit computing cost has been reduced.

The usage at MIT by Harvard users amounts to something between 10,000 and 15,000 jobs a month, a substantial level. The usage on the First Data system amounts to between 3,000 and 6,000 terminal connect hours a month, also a significant level. These partial usage statistics indicate Harvard's current commitment to remote computing services, not networks in the strictest sense but certainly in a sense that bridges easily to a network usage philosophy.

What of the future? Wearing the hats of both an administrator and a technologist, I am accustomed to living on the horns of a dilemma. The administrator keeps yelling for long-range plans, and the technologist keeps saying, "But it's very difficult to plan with a highly evolutionary technology." Both are correct. One of the things I believe very strongly, however, is that planning is a necessary continuum of activity. It is not sufficient to develop a five-year plan and then five years later look at it again. In fact, it is useful to look at it more than once a year. There are a significant number of changes that can affect planning, and often these changes are entirely unpredictable. In addition, I find that planning relative to computing activities is useless, if not impossible, unless criteria and constraints are applied that are consistent with the philosophy and organization of the institution involved. I say this because I would not vacuously try to generalize that what is good for Harvard is necessarily good for all other postsecondary educational institutions. But I do think that any institution should consider the kind of computing service philosophy that is working at Harvard.

My primary point regarding planning is that I see a significant need for a computing service methodology by which a broad taxonomy of users can be mapped into multiple "remote" computing resources in accordance with some variable set of optimization criteria.

At present, the objectives of such a methodology are perceived to include:

1. Interface capabilities to allow for the matching of classes of usage required to classes of resources available, for example, a table-driven "translation algorithm" that would be capable of transforming job control statements to match particular computer resources in the network and the like.

2. Information communication resources to provide different classes of users with a set of consultation and training modules for using of various resources such as, remote computing videotape cassettes, programmed examples, and texts.

3. A data communication resource to provide for the usage (and the user) transparent connection of local terminal facilities to various types of communications systems, for example, packet-switched networks, circuit-switched networks, private microwave, and TELCO voice grade lines.

4. A comprehensive accounting and billing system for multiple user accounts utilizing multiple resources and services.

As an overall objective, the value added by such a methodology should equal or exceed the costs. One of the more interesting and vital elements of the planning study is the determination of the economic viability of such a methodology.

The major functional relationships involved here are shown in Figure 33.1. There is a user population, which needs to be examined carefully, both independently and in conjunction with a usage population, which also needs to be examined carefully. Obviously, the two are closely related, but there is both a one-to-many mapping and, in some cases, a many-to-one mapping between the two populations. One user can characterize himself by means of any number of different types of usage. Usage, as I define it here, relates more closely to the data communication and terminal facilities required for access to computing resources as shown in Figure 33.2.

Another very significant factor in the usage of remote computing resources in general, and networks in particular, is what I have chosen to call information communication facilities. It involves getting to the user the information that is necessary for him to make adequate use of the computing resources that are available, plus getting user feedback to the computing resources to indicate how well his needs are being met. From the point of view of managing a computer resource, information communication is a significant and often neglected problem. If we consider that, for example, Harvard attempts to make use of the 360/91 at UCLA, the problem is complicated significantly by distance. It is not easy or inexpensive for a user to pick up the telephone and call the systems programmer at UCLA about some particular phenomenon

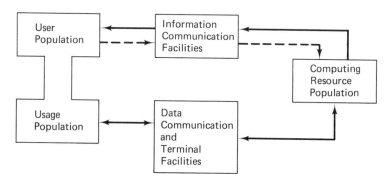

Figure 33.1. Functional relationships of the key elements.

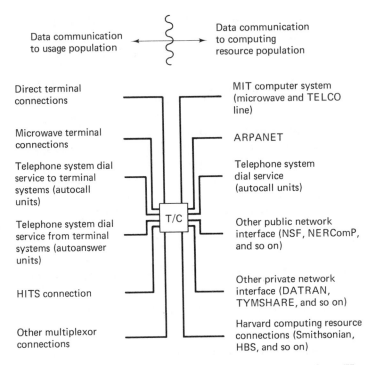

Figure 33.2. Potential translation/communication system interfaces (Harvard).

that has occurred. We are particularly interested in the role technology can play in this information communication process.

Returning to the four major objectives of the proposed methodology, I think there is both an opportunity and a great need to study seriously how to provide "automated" interface capabilities to allow the matching of classes of computing usage to classes of computing resources available. I choose, for one example, a table-driven translation algorithm to do this in accordance with variable optimization criteria. This approach may seem quite inadequate, but my initial goals are very modest. For example, it seems to me not only possible but also useful to develop such a translation algorithm to allow a Harvard user more or less automatically to switch a job from the IBM 360/91 at UCLA to the IBM 360/91 at Columbia or to the IBM 370/165 at MIT. There are, of course, significant counterexamples to a perfect solution. But I think that such a methodology will play a significant role in the usability of remote computing facilities via networks.

As for the information communication resource, several technologies are coming along that I believe will provide the needed vehicle. Easy-to-use videotape cassettes are available now at a relatively low cost. Through well-done instructional videotape cassettes keyed to programmed examples and improved handbooks, one could develop modules by which a user could pace himself, depending on his ability and interest, in learning to use a remote computing facility. There is the additional problem of continuing consultation after actual usage begins. Perhaps these technologies, in addition to the communications facilities of the network, can be brought to bear on this activity. If one looks at what really happens at many university computing centers around the country, the relationship between the user and the resource is already very remote. There are, in fact, very few one-on-one consultations. One reason, of course, is the cost of providing such services on a large scale. We now must formally address the provision of mechanisms and techniques by which this information can be disseminated. I think that in order for networks really to succeed, it will be absolutely necessary.

The third element of the methodology is a data communication resource to provide the transparent connection of local terminal facilities to multiple types of communications systems (Figure 33.2). At Harvard, for example, we already have several data communication facilities in use. The data communication with the MIT computer system is accommodated by a fifty-kilobit TELCO line. A fifty-kilobit microwave link serves for backup. We have the ARPANET interface. We have access to the computing facility at Columbia by means of voice-grade 2400-baud TELCO service. These three communications facilities are not compatible, and a nontrivial switching process must occur independent of the computing resource if a user decides to use all three facilities. With packet-switching data communications systems developing, with circuit switching not being replaced, with private microwave and other types of data communication networks being developed, and with TELCO voice-grade lines and on-campus conventions of one sort or another, I think it will not only be useful but probably necessary to develop some sort of common interface facility that will enable a user's remote terminal, transparent to the user if possible, to be connected to various data communication systems. I think that the hardware technology is available now, and my own guess is that it will be a very rewarding area in which to work to derive rather quickly if not a complete solution, at least a very useful one.

The final element of the methodology, which may sound mundane relative to the first three, is a comprehensive accounting and billing system to handle the multiple combinations of users, usage, communications, resources, and services. Such a service will probably be relatively simple to provide in terms of the techniques required. It is, however, a significant service. It is mind boggling to visualize each user having to deal separately with his own accounting. Moreover, significant data for the evaluation of economic viability could be captured and systematically evaluated through the system.

Finally, I am obliged to raise the question: Will the methodology wash? I think the answer depends on several things. One is our ability to develop useful and usable taxonomies of the type that I have been describing. Another is the value added by the methodology versus the economic and organizational realities of utilization. That is, what institutions are really in a position, economically and organizationally, to make use of networks and facilities of this type? How generalizable is the methodology? What is the long-term relationship between the cost of communication services and the predicted significant reduction in cost of computer processors and storage? One prophet has predicted a reduction by a factor of 20 in computer processor and storage prices within seven years as a result of large-scale integration in integrated circuit technology. Finally, a great deal also depends on our ability to develop successful management techniques for these networks.

Relative to management, there is one contribution from archy that I wanted to share, particularly with those who are computing center directors. This is a little poem that has been slightly modified from Marquis's original. It says,

there is bound to be a certain amount
  of
trouble running any technological resource
if you are its director the trouble happens
  to you
but if you are a dealer you can arrange
  things so
that most of the trouble happens to
  other people

So, consider being a dealer as well as a director.

References

1. Lines of poetry from *ARCHY AND MEHITABEL* by Don Marquis. Copyright 1927 by Doubleday & Company, Inc., Garden City, N.Y. Reprinted by permission of the publisher.

2. John A. Austin, "Planning Computer Services for a Complex Environment," *Proceedings of the 1971 AFIPS Fall Joint Computer Conference*, AFIPS Press, Montvale, N.J., 1971, p. 541.

34

GERARD P. WEEG
University of Iowa

Regional Star Networks
as Seen by the User
and Server*

Should a national network of interconnected large, independent, free-standing, and equal computers appear in this country in the near future, increased potential would be provided to the giant centers without a doubt. But that would be a case of the rich once again getting richer. My concern is for the 1,126 colleges out of a total of 2,807 institutions of higher learning in this country which, as late as 1970, were without access to computing of any sort. With the current breakthrough on minicomputers, this number will no doubt be cut severely. But I fear that we will then see islands of relative ignorance not sharing in the capabilities and experience of the major centers. Star networks can be the capillaries leading out from the arteries of the national network, providing computing power to the small colleges and even the secondary schools of the nation. The need for this system remains despite the rise of minicomputers.

At present there are essentially two kinds of star networks: those depending primarily on typewriterlike terminals and using either a time-sharing system as at Dartmouth or a remote job entry environment as at the University of Texas, and those depending upon remote batch terminals. Only rarely was a network cast in one or the other of these forms by design. Usually it assumed the form the central facility was already prepared to accept.

The intention, at least of all of the NSF-sponsored star networks, was to provide computing power to smaller institutions so that they could make rapid progress in introducing computing into their instructional process. The hope was that the smaller institutions could have easy access to huge computers, large libraries of programs, well-staffed computer organizations, and even cheaper computing. To a large degree all these hopes were realized, though it is only honest to point out that about half the star networks have failed or atrophied. It is equally honest to point

* The figure and all tables in this chapter were originally published in 1973 by the University of Iowa, Iowa City, in *A Study of University Networks* by Fred Weingarten, Norman Nielsen, James Whiteley, and Gerard Weeg. The material is used with the kind permission of the University of Iowa.

out that the survivors are growing larger and stronger, and with the natural pressure of state legislatures the number of star networks is increasing.

Star networks commonly display several features: commitment to success of both server and user; financial surprises; staff and faculty involvement; hardware, software, communications, and documentation pressures; curriculum revisions and the paucity of good material; administrative data processing producing unexpected benefits; and the natural empire-building effect. And in a separate category is the degree of green wood left in the college faculty.

The issue of commitment is of primary importance on both ends of the line. Given adequate commitment, no set of negative circumstances will destroy a network. Lukewarm commitment at a remote institution will cause the loss of that school, and at the central facility it will cause the decline and demise of the network. What is this commitment? At the small-college end, it involves both the administrative structure and the faculty. The administrative part includes particularly the president or academic dean and the business officer. If one of these two agencies opposes the network, it will fail; if both give it strong support, it will be hard for the faculty to escape the use of the computer. The direct and subtle pressure of the administration to get with computing clearly motivates the faculty to do so. If there is a cadre of faculty members who are willing to try something new with computing, success is much closer.

Financial surprises occur at both ends of the line. The first fact to become obvious to the remote college is that fairly substantial sums of money are leaving the campus to pay for computing. What is even more horrifying is that the ratio of dollars spent at the central facility on actual computing to dollars spent on communications, terminals, and local staffing is remarkably low, typically on the order of 1 to 10. Finances at the central facility are even more of a surprise, and here is where commitment, and perhaps sanity, is most sorely tested. After four or five years of involvement, the central facility's expenditures to support a star network usually slightly exceed or about equal the income generated. It soon realizes, however, that the balance of payments can easily be brought back into a good shape simply by increasing the number of remote nodes. This latter is true usually because neither central facility staff nor hardware would have to be augmented to handle up to nearly twice the number of remote customers.

It was learned early that the staff at the central facility needed to serve the remote user is of a different kind and number than was anticipated before the network got under way. A full-time network coordinator is absolutely necessary at the central facility. His role is part technical, part promotional, part creative, and certainly entrepreneurial. Working with him will usually be at least one person dedicated to computer-oriented curriculum development. There will also be one or two programmer-analysts who will double as long-distance-telephone computer debuggers and as circuit riders. In short, the central facility staff dedicated to the network can easily mount up to four or five people, with substantial slices of assistance from the director and the technical staff.

At the remote end, it also became clear early that each campus needs a faculty campus coordinator. This man need not have started with computer expertise. Rather his principal qualifications are willingness to learn, ability to communicate with and motivate other faculty members, and ability to discern areas of possible computer uses. This person is a key to success or failure on a campus. Further, release of less than half his time to this duty is equivalent to zero released time. Other staff implications on the remote campus follow. A work-study student or two will generally become hot-shot local programmers and operators at the batch sites. If administrative data processing is to be pursued on the terminal, a half-time person with broad experience in the business or registrar's office must be assigned to help make it work.

As far as hardware and software are concerned, they must be provided in sufficient quantity (a single typewriter is an abomination), quality, and pertinence to local needs. They must work as advertised from the first day; they must be documented (that is, simple cookbooks must accompany them); and when the inevitable failures occur, succor must be immediate. These requirements have a vast implication for the central facility. The communications-oriented hardware probably has to be acquired, and the myriad of communication devices must be sorted out into the good and the "never touch." Arousing the several phone companies with which one works to learn about data transmission and to give prompt repairs is no longer the problem it was four or five years ago, but it is still at least 10 percent of the central coordinator's job. The provision of documentation to the remote school is a story in itself. The type of documentation available at the normal center is comprehensible to only the cognoscenti. Simple, step-by-step documentation is needed for the new schools.

Since the goal of most networks is to change and, we hope, to improve education, the central facility must work with the faculty of the remote institutions (and for that matter, with faculties at other star networks) to produce new written course material incorporating computer usage in such a way as to improve the course. This job is the most monumental task facing us, and it will be with us until our systems of rewards and punishments are appropriately adjusted to provide for it.

At least with batch networks, and increasingly with typewriter networks, cooperative efforts to produce ADP packages have been surprisingly successful. At Iowa, for example, we have cooperatively built systems for registration, grade reporting, and allied topics that allow the remote school to maintain and provide student records for about a dollar a student a year. The design and programming effort that goes into designing and programming these systems, however, is enormous.

The empire effect works two ways. It causes the central facility to want to gobble up more schools, and it causes remote schools to want to sever their ties and have their own little center. No cures or preventatives are known for this malady.

Green wood left in the remote school is the crux of the matter. The degree to which the faculty is loaded up with class time or is committed to preserving the present curriculum determines success or failure to a large extent.

Abundant statistics accompany this chapter to provide details on the Iowa star network (Figure 34.1, Tables 34.1, and 34.2) and other systems. They will flesh out many of the things set out in skeleton form in this discussion.

The costs of operating the RCC (Regional Computer Centers) are all passed back to the RCC schools, with the exception of the salaries of the central staff members dedicated to the RCC, which are reflected into the general overhead of the Center. In early 1972, this staff commitment was as follows:

UCC Director	5%
RCC Coordinator	100%
(C. Shomper)	
Programmers	200%
Secretary	50%
Total annual salary	$54,000

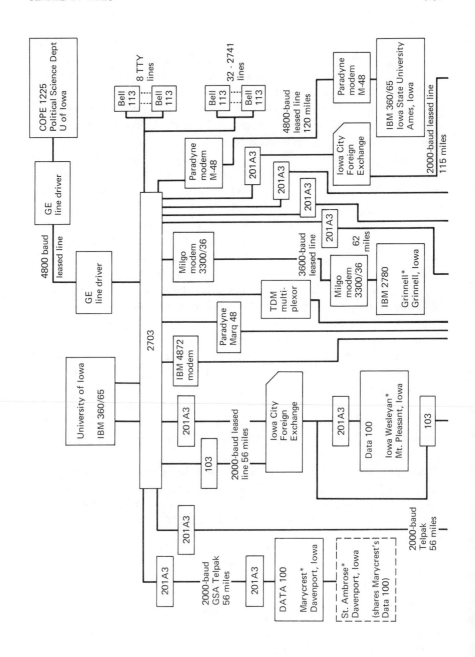

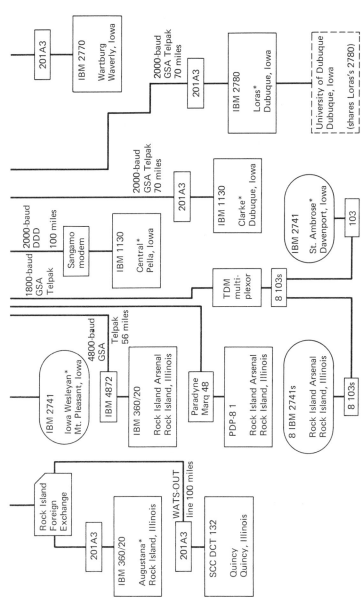

* Indicates that these schools have received NSF support for network participation.

The operating system used is OS-MVT-version 20.1-HASP 3.0. The typewriterlike devices use ATS, CPS, and their associated RJE facilities, in addition to Coursewriter.

Figure 34.1. The Iowa Regional Computer Center Network.

The costs of the RCC as the participating colleges perceive them in the last complete year (July 1, 1970—June 31, 1971) are given in Table 34.2. RCC services and usage are summarized in Tables 34.3 and 34.4.

Table 34.1. 1968 Regional Computer Center at the University of Iowa

College	Public or Private	Full-Time Equivalent Enrollment	Full-Time Equivalent Faculty	Total Operating Budget
University of Iowa	Public	20,000	1,100	$100,000,000
Iowa Wesleyan College	Private	825	53	2,500,000
Kirkwood Community College	Public	1,279	120	4,700,000
St. Ambrose College	Private	1,270	90	3,000,000
Central College	Private	1,285	90	3,200,000
Clarke College	Private	982	105	3,000,000
Coe College	Private	901	82	3,700,000
Marycrest College	Private	1,100	66	1,700,000
Loras College	Private	1,460	119	4,300,000
Grinnell College	Private	1,056	122	5,600,000
Augustana College	Private	1,896	104	5,900,000
University of Dubuque	Private	1,048	63	2,600,000

Table 34.2. Cost of the Regional Computer Centers Participating July 1, 1970—June 31, 1971

	Terminal Type	Communications Type	IBM 029 Key Punches	Annual Equip. Cost*
Iowa Wesleyan	DATA 100 $832/month	Foreign exchange $350/month	2 $140	$12,960 12,960
St. Ambrose	Share with Marycrest		2 $140	10,000 10,000
Central	IBM 1130 $504/month†	WATS dial up $302/month	3 $250	9,050 9,050
Clarke	IBM 1130 $100/month†	TELPAK $226/month	2 $80	1,860 1,860 (maintenance only)
Marycrest	IBM 2780	TELPAK $200/month	3 $210	11,991 10,585
Loras	IBM 2780 $344/month	TELPAK $80/month	2 $138	5,800 5,800
Grinnell	IBM 2780 $982/month	Leased line $487/month	6 $378	16,320 16,320
Augustana	IBM 360/20 $936/month	Foreign exch. and TELPAK $166	3 $210	13,681 13,781
U. of Dubuque	Shared with Loras $638/mo.	$148/month	1 $70	8,496 8,496
Wartburg	IBM 2770	WATS dial up	2	8,466 8,863
Quincy	SCC 132	WATS, foreign exch., and TELPAK $847	3 $208	29,384 (Purchase) 6,635

* The second figure is the amount budgeted for FY1971–72.
† This is the portion of the cost assigned to the 1130 as a terminal.

Table 34.2. (Continued)

	Annual Comm. Cost*	Salaries		Computer Time*	Misc.*	Total*
Iowa	$4,560	1.5FTE	$12,000	$5,933	$2,200	$37,653
Wesleyan	4,560	1.5	12,000	6,500	2,200	38,220
St.	1,100	1.25FTE	9,660	2,800	1,275	24,835
Ambrose	1,250	1.25	9,660	3,200	1,275	25,385
Central	4,361	2FTE	12,140	1,963	1,325	28,839
	3,624	2	13,359	2,500	1,800	30,333
Clarke	2,743	1-1/3FTE	11,875	3,935	2,300	22,713
	2,743	1-1/3FTE	11,875	4,000 (est.)	2,300	22,778
Marycrest	2,100	2FTE	13,800	4,567	1,000	33,458
	2,100	2.5FTE	17,450	3,000	1,000	34,135
Loras	960	1.25FTE	4,415	1,902	359	13,436
	960	1.25	5,500	2,000	4,840	19,100
Grinnell	5,897	3.25FTE	28,000	10,421	2,173	62,811
	5,906	3.25	29,000	8,250	1,651	61,127
Augustana	3,620	.75FTE	4,012	3,961	607	25,881
	2,000	.75	4,012	1,600	607	22,000
U. of	1,776	.25FTE	760	3,046	260	14,338
Dubuque	1,776	.25	760	2,308	1,432	14,772
Wartburg	2,961	6/7FTE	7,631	644	51	19,753
	3,480	1.5	7,820	2,200	1,240	23,603
Quincy	9,115	2	1,929	149	688	41,265
	8,263	2	2,592	5,763	1,107	24,360

* The second figure is the amount budgeted for FY1971–72.
† This is the portion of the cost assigned to the 1130 as a terminal.

Table 34.3. Services for the Participating Regional Computer Centers

	Operating Hours Weekdays	Weekend	Terminal Space
Iowa Wesleyan	7 A.M.– 9:30 P.M.	None	400 sq ft
St. Ambrose	same as Marycrest		none
Central	11 A.M.– 10 P.M.	Sat 11 A.M.– 2 P.M. Sun 7 P.M.– 10 P.M.	550
Clarke	7 A.M.–12 midnight	7 A.M.–12 midnight	450
Marycrest	8 A.M.–5 P.M.	None	200
Loras	9 A.M.–6 P.M.	Sat 6 P.M.– 8 P.M. Sun 1 P.M.– 3 P.M.	500
Grinnell	8 A.M.– 5:30 P.M. 7 P.M.–9 P.M.	Sat 1 P.M.– 3 P.M. Sun 7 P.M.– 9 P.M.	270
Augustana	10–10:15 A.M.– 2–2:15 P.M. 6 P.M.–10 P.M.	none	225
U. of Dubuque	same as Loras		
Wartburg	8 A.M.–4 P.M. 7 P.M.–9 P.M.	1 P.M.–4 P.M.	500
Quincy	10 A.M.–2 A.M.	2 P.M.– midnight	100

Table 34.3. (Continued)

	User Space	Open/Closed Shop	Assistance to User Who Provides It	
Iowa Wesleyan	750 sq ft	open	RCC coord. students	½ FTE 1 FTE
St. Ambrose				
Central	500	semiclosed	RCC coord. students	¼ FTE ½ FTE
Clarke	450	semiclosed	RCC coord. faculty & students	2 FTE
Marycrest	315	closed	RCC coord. Staff	½ FTE 2 FTE
Loras	600	closed	RCC coord. students	½ FTE ½ FTE
Grinnell	426	semiclosed	Faculty & staff students	3-½ FTE ½
Augustana	800	closed	RCC coord. & Students	¼ FTE
U. of Dubuque				
Wartburg	500	semiclosed	Instructor & assistant	
Quincy	400	closed	RCC coord. & student	

Table 34.4. Regional Computer Center Usage (UCC Usage Excluded) July 1, 1970–June 31, 1971

	Cards In	Cards Out	Pages Printed	Lines Printed	CPU Seconds	Elapsed Seconds
Grinnell	2562818	24330	105072	3414261	51911	135258
Marycrest	925540	326	146615	1750557	21521	70852
St. Ambrose	308073	0	27338	734794	11273	41441
Augustana	695855	7547	42506	1518406	18536	65282
Iowa Wesleyan	422990	254	41730	1336434	21005	90705
Central	183081	3109	11232	329999	9852	24259
Clarke	289470	644	16508	515952	15401	74896
Loras	369877	1695	20849	592534	9136	35047
U. of Dubuque	177141	38	25841	750521	11127	37182
Wartburg	156212	43	5885	209562	2963	12409
Quincy	44250	280	4965	60304	781	2362

Table 34.4. (Continued)

RCC Total	6135307	38266	448541	11213324	173513	589697
Grand Total at the U. of Iowa	91644384	5631170	7299410	250556520	6918477	13599020
%RCC of total	6.69%	0.68%	6.14%	4.48%	2.51%	4.34%

	Fast Core Megabyte Seconds	Large Core Megabyte Seconds	Magnetic Tape Device Seconds	Disc I/O Accesses	$ Charge	$ Jobs
Grinnell	16297	82	16507	344503	10421	7860
Marycrest	7189	1	775	231453	4567	3067
St. Ambrose	5035	965	990	119722	2800	2966
Augustana	6035	1978	1504	73808	3961	4358
Iowa Wesleyan	10879	1696	5338	933364	5933	5791
Central	2811	0	4487	722259	1963	1198
Clarke	7980	3	832	220556	3935	2924
Loras	3070	1	916	70843	1902	3229
U. of Dubuque	5772	0	12713	310307	3046	1855
Wartburg	1060	0	0	45823	644	975
Quincy	207	0	0	0	149	191
RCC Total	66339	4730	44062	3072638	39326	34414
Grand Total at the U. of Iowa	1765125	280833	4269630	105040505	1353214	350500
%RCC of total	3.76%	1.68%	1.03%	2.93%	2.91%	9.8%

The average speed of the terminals used at these colleges as measured in February/March of 1972 was 249 cards per minute and 132 lines per minute. Some average figures are then computed and presented in Tables 34.5 and 34.6.

Table 34.5. Statistics for the RCC Colleges*

Average number of students	1,200 students
Average annual budget	$3,640,000
Average card read speed	249 cards/minute
Average line printer speed	132 lines/minute
Total RCC* cards read 70–71	6,091,057 cards
Total RCC* lines printed 70–71	11,153,020 lines
Total RCC* computer charges 70–71	$39,177
Total RCC* jobs 70–71	34,223 jobs
Cards read per average college	609,000 cards
Lines printed per average college	1,115,000 lines
Number of jobs per average college	3,422 jobs
Average transmission hours per year	182 connect hours
Average computer cost/year	$3,900
Average RCC school cost/year	$28,400
RCC school cost/year range	$13,400 to $62,800
Average cost/job	$8.30
Average cost† of computing/student	$23.67

* Quincy College is excluded from these averages because of its unique role in the RCC.
† This is the result of dividing the total number of students (1,220) into the total computing cost for the average school ($28,400).

Let us consider an average 1,000-student college. It might be considered to be significant computer use if one-fourth of these students each received one assignment per week requiring the use of a computer. The significance of use is assumed to increase if the number of assignments per week increases or if the number of students involved increases. Tables 34.7 to 34.9 hypothesize what the impact of varying degrees of instructional computer use would be. The bases for these tables are the Iowa, Texas, and Dartmouth networks. These three networks represent remote batch; remote job entry via teletype to a batch system; and remote interactive computing, respectively. Clearly, some of the numbers are sheer conjecture. As the amount of computing increases, the cost of computing as such can decrease, although certain other items, such as terminal cost, will not decrease. The conclusion is that the exact impact as stated in the upper ends of the tables tends to be more untrustworthy than that at the lower end.

Table 34.6. Total Costs to Members of the Dartmouth College Consortium FY 1970–1971

College	Enrollment	Kind and Number of Terminals	Annual Cost of Terminals	Cost of Staff*	Communications	Connect Charge	Total Annual Cost
Bates	1,050	4 ASR 33	$3,120	0	$4,280	$24,000	$31,400
Colby Junior	600	1 ASR 33	780	0	560	6,000	7,340
Middlebury	1,600	3 ASR 33	1,900	0	2,450	20,000	24,350
Mount Holyoke	1,800	3 ASR 33	900†	0	4,300	18,000	23,200
New England	1,000	1 ASR 33	300†	0	590	6,000	6,890
Norwich U.	1,200	1 ASR 33	780	0	1,448	6,000	8,228
Vermont Technical	450	1 ASR 35 1 KSR 35	1,320	0	1,448	12,000	14,768
Average College‡	1,100	2 ASR 33	1,560	0	2,154	12,000	15,900

* There is no general released time for staff in the Consortium Colleges.
† These two colleges own their terminals. These costs are the estimated amortization for the terminals based on a purchase price of $1,200 and a four-year period.
‡ The annual cost of terminals is based on $780 rented for one terminal. Communications is the average assuming 15 terminals, since the KSR 35 is connected to the system. Similarly for the connect charge.

Table 34.7. Impact of Significant Instructional Use Iowa RCC

No. of Students Computing	Transmission Time, Hours/Week (58 cards, 93 lines/run) (2.81 min/3 runs)	Computer Cost per year, 40 Weeks per Year (7 cents/run)	Communication Cost per Year ($3,600)
One assignment per week, 3 runs per week per student			
250	11.70 hr/wk	$ 2,100	$3,600
500	23.40	$ 4,200	
750	35.10	$ 6,300	
1,000	46.80	$ 8,400	
Two assignments			
250	23.40	$ 4,200	$3,600
500	46.80	$ 8,400	
750	70.20	$12,600	
1,000	93.60*	$16,800	
Three assignments			
250	35.10	$ 6,300	$3,600
500	70.20	$12,600	
750	105.30*	$18,900	
1,000	140.40*	$25,200	

* Turnaround would become a serious problem, and faster communication lines would be needed. This would increase the total cost by about $5,000 per year.

Terminal Cost per Year ($10,800)	Keypunch Cost per Year (3 per Assignment per 250 students)	Salaries and Miscellaneous ($10,712/yr)	Total Cost per Year
One assignment per week, 3 runs per week per student			
$10,800	3-$2,520	$10,712	$29,732
	6-$5,040		$34,352
	9-$7,560		$38,972
	12-$10,080		$43,592

Table 34.7. (Continued)

Two assignments			
$10,800	6-$5,040	$10,712	$34,352
	12-$10,080		$43,592
	18-$15,120		$52,832
	24-$20,160		$62,072
Three assignments			
$10,800	9-$7,560	$10,712	$39,972
	18-$15,120		$52,832
	27-$22,630		$66,642
	36-$30,240		$80,552

Table 34.8. Impact of Significant Instructional Use DARTMOUTH NETWORK

No. of Students Computing	Connect Hours/ Week (21.1 min. or .35 hour/job)	No. TTY's Needed (30 hours/TTY)	CPU Cost $500/ Month per Port, 2 TTY's/Port (9 mos)
One assignment per week, 2 sessions/ assignment			
250	175	6	$ 13,500
500	350	12	27,000
750	525	18	40,500
1,000	700	24	54,000
Two assignments			
250	350	12	27,000
500	700	24	54,000
750	1,050	36	81,000
1,000	1,400	48	108,000
Three assignments			
250	525	18	40,500
500	1,050	36	81,000
750	1,575	54	121,500
1,000	2,100	72	162,000

Table 34.8. (Continued)

Communications Cost per Year $1,000/port	TTY Lease $600/Year/TTY	Total Cost per Year
One assignment per week, 2 sessions/ assignment		
$ 3,000	$ 3,600	$ 20,100
6,000	7,200	40,200
9,000	10,800	60,300
12,000	14,400	80,400
Two assignments		
6,000	7,200	40,200
12,000	14,400	80,400
18,000	21,600	120,600
24,000	28,800	160,800
Three assignments		
9,000	10,800	60,300
18,000	21,600	120,600
27,000	32,400	180,900
36,000	43,200	241,200

Table 34.9. Impact of Significant Instructional Use TEXAS NETWORK

No. of Students Computing	Connect Hours/ Week (0.19 hour/job)	No. TTY's Needed (30 hours/TTY)	CPU Cost $.40/Job per Year/ 40 weeks	Connect Change $1/Connect Hours per Year
One assignment per week, 2 jobs/assignment				
250	95	3	$ 1,520	$ 3,800
500	190	6	3,040	7,600
750	285	9	4,560	11,400
1,000	380	12	6,080	15,200

Table 34.9. (Continued)

Two assignments 2 jobs each				
250	190	6	3,040	7,600
500	380	12	6,080	15,200
750	570	18	9,120	22,800
1,000	760	24	12,160	30,400
Three assignments 2 jobs each				
250	285	9	4,560	11,400
500	570	18	9,120	22,800
750	855	27	13,680	34,200
1,000	1,140	36	18,240	45,600

Communications $600/Year/TTY	TTY Lease $600/Year/TTY	Fixed Cost $8,700 (salaries and supplies)	Total Cost per Year
One assignment per week, 2 jobs/assignment			
$ 1,800	$ 1,800	$8,700	$ 17,620
3,600	3,600		26,540
5,400	5,400		35,460
7,200	7,200		42,880
Two assignments 2 jobs each			
3,600	3,600	$8,700	26,540
7,200	7,200		44,380
10,800	10,800		62,220
14,400	14,400		80,060
Three assignments 2 jobs each			
5,400	5,400	$8,700	35,460
10,800	10,800		62,220
16,200	16,200		88,980
21,600	21,600		115,740

35

DAVIS B. MC CARN
The National Library of Medicine

A Medical
Information Network
and Constraints
on Networking*

First, I shall describe the on-line medical information network of the National Library of Medicine (NLM), and then I shall speculate on how the situation may be different in higher education, and on why it may be much more difficult there to establish on-line networks.

The National Library of Medicine provides bibliographic services to the nation's medical community to assist in translating research into improved health care, to improve access to the literature for continuing and undergraduate medical education, and to improve communication between health researchers. The computers at the National Library of Medicine are used to produce the *Index Medicus,* the comprehensive, internationally used index to the world's biomedical literature. This index and a variety of special bibliographies are produced by the Medical Literature Analysis and Retrieval System (MEDLARS). One of the computers at the NLM was connected in October 1971 to a nationwide data communications network and provides an on-line bibliographic search service, MEDLINE (MEDLARS On-line), for medical schools, hospitals, and research institutions throughout the country. In addition, the data base for an on-line toxicology information service is produced.

In 1973 the MEDLARS system provided bibliographies from a base with an input of more than 220,000 citations during the year. The total file of citations to the medical literature has now grown to 1.7 million. MEDLINE now allows almost instantaneous, interactive searching of more than 400,000 citations from the world's significant biomedical serial literature. One year from its birth the service was supporting an average of 25 simultaneous users, 46 hours per week. In December 1972, 10,605 searches were processed—a rate of nearly 140,000 searches per year.

Service is provided through a data communication network that allows access through a local dataphone call in any of 41 metropolitan areas across the nation. More than 130 institutions with more than 200 terminals are using the service. The communications network is also being used on a trial basis by the French MEDLARS Center, and the Canadian

* This paper is based on work conducted by the National Library of Medicine and thus is not subject to copyright.

MEDLARS Center is a regular part of the network. The MEDLINE data base is also operated from a computer in Sweden and is connected by remote terminals in six locations in Scandinavia through regular telephone lines. MEDLINE is the first generally accessible on-line, interactive information service. It constitutes the first national and internation "hard-wired" science information network. The library plans to expand service to every state in the nation. The number of using institutions will be increased to 350. In this expansion, emphasis is being placed on health care delivery.

The MEDLINE service is an outgrowth of an experiment that provided an on-line search service against the journals in the *Abridged Index Medicus* (100 of the top journals in clinical medicine) via the Teletypewriter Exchange Network (TWX) of Western Union. This service, AIM-TWX, was initiated in May 1970 and drew such an immediate response from the medical community that the library began the planning for a service to be based on its own computer in late 1970.

This planning had, as one of its primary concerns, the development of an adequate communications network. Initial analyses indicated that two-thirds of the total cost of using such a service would be telephone costs if the AT&T direct dial network were used. The National Bureau of Standards assisted NLM and conducted a cost analysis of alternatives; it was concluded that a competitive procurement might result in a major savings in communication costs. The NLM issued a request for proposals in June 1971 and obtained responses from only two vendors, Western Union and TYMSHARE, Inc. The library contracted with both for communications services. Western Union Datacom service was installed in six cities in November 1971, and the NLM computer was connected to the TYMSHARE network in February 1972.

The Western Union Datacom service, however, forced NLM to become the center of a star network, whereas the TYMSHARE network allows the library to be one host on a distributive network. The latter position demonstrated such obvious advantages that the Western Union service was dropped in November 1972. The library now relies on the TYMSHARE network, direct connections to the direct dial network, the TWX network, and the Federal Telecommunications System.

The ARPA network was seriously considered but was finally rejected because it was then highly experimental, it did not provide adequate geographic coverage, and it would have involved access for MEDLINE users through the various university host computers. NLM was seeking only a distributed network to connect terminals to a variety of computers, not a computer resource-sharing network, and the latter did not appear adequate to meet the requirements of the former.

The National Library of Medicine now supports six computer services on the TYMSHARE network: MEDLINE on the NLM and on the SDC computers; the Toxicology Information System (TOXICON) on the computer of COMNET, Inc.; computer systems to support clinical decisions at Massachusetts General Hospital; computer-aided instruction in undergraduate medical education at Ohio State University; and a simulated patient encounter (CASE) on the computer of the University of Illinois.

Networking has been contagious, and several commercial firms are connecting services to the same network. The Educational Resources Information Center (ERIC) base, the National Technical Information Service (NTIS) base, and *Chemical Abstracts Condensates* are all available for on-line searching. At present, there are nearly 1.5 million citations or abstracts available for on-line search through the TYMSHARE network.

Network services from TYMSHARE are provided on a "resource pricing" basis; that is, the major cost is the amount of information transmitted, not the time of connection. Actual charges are (1) $.40 per log-in, (2) $1.00 per connect hour, and (3) $.125 per 1,000 characters transmitted. The cost of communications is about $4.90 per hour on the MEDLINE system; dial telephone costs would be about $24.00 per hour.

Constraints on Networking

In observing the development of the MEDLINE service, I have been struck by how much easier it has been to develop an on-line user-oriented network in the medical arena than it appears to be in higher education. This has led me to wonder about the characteristics of computing in higher education that might make the problems of networking particularly difficult there. I have arrived at the following tentative conclusions.*

Several different types of networks now exist or are under development: those for data transmission, those for sharing computer resources, and those for gaining access to computers. The data transmission networks are generally designed to handle large volumes at high rates. The network of Datran appears to be of this type, as is the Advanced Record System of the General Services Administration. The ARPA network and MERIT

---

* The opinions expressed here are solely those of the author and do not reflect policies of the National Library of Medicine or the federal government.

are examples of computer resource-sharing networks. Access to such networks is generally through the host computers on the network; an exception is the recent development of the terminal interface processor (TIP) for the ARPA network. A computer access network is different from either in that the network and access thereto is independent of the host computers; any terminal can call the network and have access to any host that has appropriate access codes. Users are independent of any particular local computer system.

Since networking in a real, hard-wire sense is not complex or difficult, why has higher education made such modest progress in developing networks? The answer is probably rooted in the unconscious attitudes of the participants, attitudes that have gone unchallenged by the users. Put oversimply, university computing is a cottage industry. The integrity of the market in each school depends on keeping the users as a captive market of their computer centers rather than on networking.

This interpretation seems consistent with what has been accomplished. Clearly, administrative networking poses no threat; exchange of programs and data bases increases the use of local computers. Such exchange arrangements do not work well because the provider of the program or data base bears the substantial cost of making it communicable, but at least it can be called networking. Even the development of such real networking as does exist is profoundly affected by the cottage industry mentality. Both the ARPA (with the exception of a few TIPS) and MERIT networks require the user to enter the network through a host computer. Thus, the local market is kept dependent on local computers, and this is justified in terms of the dubious future possibility of having programs in one computer communicate with programs or data bases in another.

This situation seems to be an inevitable result of the economics of providing information and computer services. The most important characteristic of information services to discuss is their cost. Analysis indicates that the cost per unit service, that is per search, decreases as the numbers of searches increases. Mathematically,

$$\frac{C(N_1)}{N_1} < \frac{C(N_2)}{N_2}$$

if $N_1 > N_2,$

where N equals the number of searches and $C(N)$ equals the cost of providing N searches.

This characteristic is profoundly important because it leads to the following situation:

$$\frac{C(N)}{N} > C(N + 1) - C(N).$$

That is, the marginal cost of the $N + 1^{\text{th}}$ search is less than the average cost. In a free market, the price paid for an item seeks an equilibrium where the price equals the marginal cost. At this price, an information service must lose money because at all levels the marginal cost cannot recover the cost of providing the service.

This characteristic has been known in economics for many years. Such economic situations have three alternative stable situations: (1) the service can fail and not be provided at all, (2) the service can be provided by a monopoly or oligopoly for an administered market if the demand allows charging well above the marginal cost, or (3) the service can be provided or regulated by the government if the social value warrants the provision of such a service.

The viability of the second alternative depends on the elasticity of the demand curve. A demand is said to be elastic if total revenue increases per unit sale; that is,

$$N_1 P_1 > N_2 P_2$$

if

$$P_1 < P_2,$$

where N_1 equals the number of units demanded at Price P_1. If the market is elastic, revenues can be increased by lowering the price. In this situation, even a monopolistic provider of a product can improve his situation by dropping his price until the marginal return equals the marginal cost. He is still losing money, but not as much. In any case, given an elastic market, even the monopolistic supplier must go broke.

On the other hand, given an inelastic market, the monopolistic supplier may be able to set a price that recovers his costs and thus allows him to survive in the business. The price must, however, be higher than the marginal cost of the service, and thus it does not meet the optimal demand for the service.

If the service has both a price and a social value and if the price is less than the price plus the social value, then society is short-changed by the monopolistically administered price and the only optimal solution must include at least partial public funding.

It seems highly probable that the demand for information services is highly elastic over the higher price ranges. If this is true, such services will have to be, in large part, publicly funded. An attempt to recover costs may be self-defeating because costs cannot be recovered. But each increase in prices for services sharply reduces the social value of the service provided.

As Samuelson states the case: "Actually, requiring Price = (Marginal Cost) *is* the ideal target for optimal efficiency. . . . Much of monopoly, oligopoly, and other forms of imperfect competition come from decreasing-cost cases—setting $P = MC$ while AC is still falling will involve the firm in a *chronic loss*. . . . How can society achieve its ideal of $P = MC$, where the Marginal Utility of output just matches its Marginal Cost at the equilibrium amount? The answer involves a *permanent government subsidy* to the decreasing cost producer. Where does the subsidy come from? From the general budget."[1]

The foregoing considerations also indicate that commercial alternatives are not viable except in inelastic markets. As with the time-sharing business, it looks as though the economics are such that such services must be partially publicly funded or fail as commercial endeavors unless prices much higher than the marginal cost can be collected.*

The previous discussion demonstrates that decreasing cost services are in a peculiar market position; free competition would inevitably result in mass bankruptcies. In order to be viable, such services need funds to recover capital and start-up costs or enough leverage in the market to keep prices up. Control of the market can be achieved through product differentiation which, in effect, creates a minimonopoly. Market control can also be achieved through expanding in size to become large enough to administer prices and exercise restraints against price wars. One way of expanding at minimal cost is through networking. If one reviews the survey of the 28 profitable time-sharing companies (out of 150) in *Time-Sharing Today,* it is apparent that every one of these companies either differentiated and offers a unique service or has a network or

* Many markets are elastic over part of the price range and inelastic over other parts. In such a situation, providers would tend to operate in the inelastic range.

both.[2] These economics also describe the position at university computer centers and explain at least in part the lack of networking.

But networking has become a reality. As noted earlier, commercial firms are rapidly entering the information services business, and at least in this area university computers may have to network to survive. The technological alternative is clear: build a network that divorces user access from host computers so that any user can reach any computer from any network node.

This alternative may not be favored by computing center directors because it would have profound effects on university computing. First, it would allow competition; if the faculty at a school felt the service at their school was insufficient in any way, they could arrange to dial another—a frightening prospect! But even more important, it might allow the economies of the division of labor. Centers devoted to specific services could be established. Instead of accumulating many data bases on one computer, as has been the practice, and paying the high capital costs of this approach, one computer (or more) could provide retrieval services for all users. Instead of many systems playing with computer-aided instruction in a field, one (or more) centers could do the job well for many schools. Such specialization is beginning in economics, where at least two firms offer comprehensive data bases and analytic tools.

In a sense, the growing conflict is between the interests of users and those of computer science departments. If a university center were to specialize in, say, Loentiev models and related data bases, how could the computer science students get a well-rounded educational experience? The National Library of Medicine does run on-line programming (TSO) and batch processing services while providing its normal service, so this may not be a major problem. But it is an issue that already exists in another form. How can students and faculty continue to fiddle with the operating system and still provide reliable service? However these issues are resolved, the future seems to lie with such specialized services in many applications.

References

1. Paul A. Samuelson, *Economics*, 8th ed., McGraw-Hill, New York, 1970, p. 479.

2. *Time-Sharing Today*, Vol. 2, July 1, 1971, pp. 1–14.

36

RANDALL M. WHALEY
University City Science Center,
Philadelphia

Promotion and Economics of Resource Sharing

In some respects I approach this subject from a layman's point of view. Perhaps the most significant involvement I have had with computing before the recent development of UNI-COLL in Philadelphia was the decision I made when I was vice-president of Wayne State in the early sixties to promote consolidation of computing on that campus. It was then a very hot issue. Dr. Walter Hoffman, a member of my staff, was then director of the Computing Center, and I think we did a reasonably good job in improving computing through consolidation.

Let me begin with a few definitions and assertions. *I define a shared computer resource system as one that allows institutions in a region to seek the benefits of increased power and decreased cost by cooperative efforts and by multiinstitutional use of facilities and personnel.*

The first assertion is that a shared computer resource system must be planned, operated, and managed like a business. This should be true whether the shared system operates in a stand-alone mode like UNI-COLL or is attached to a single institution. The shared system encompasses hardware, software, telecommunications, and a full range of technical and management personnel. Throughout the workshop sessions so far, and throughout the presentations here today, I sense that there is support for this viewpoint as shown by the pervasive intent to try to bring cost realism into the operations of computing services.

Second, I will assert that the problems of developing, operating, and managing computing networks—however one defines network—are the same as those encountered in the development, operation, and management of shared computing resource centers.

As a final assertion, I believe that the success of a network rests upon the successful operation of strong, stable computer resource systems at the nodes of that network. I use the term network here in a generic sense, encompassing perhaps a multiplicity of interconnections among resource centers large or small throughout the nation.

A Shared Computer Resource Center as a Business Operation

Assuming problems of developing and operating networks are comparable to those in developing and operating a shared computing resource center, I will draw upon our experiences at UNI-COLL. If you look at a shared resource center as a business, in my judgment there are six important elements to be considered: (1) definition of products or services and the market, (2) creation of an appropriate business organization, (3) initial capitalization, (4) promotion and marketing, (5) production and delivery of products and services, and (6) management objectives.

We addressed all these elements as we created UNI-COLL. We have had varying degrees of success and of frustration. And we agree with Joe Wyatt of Harvard, who said, when discussing that institution's plans for computing, that planning is a continuing process. The development of a shared center like UNI-COLL so that it would be operated as a business is also a continuing process. There have been few, if any, proto-types to follow. There is a great deal to be learned and a great deal of experience to be shared in this process of developing shared resource centers and systems.

In some respects UNI-COLL is unique. We have received no state or federal subsidies in its development. We are fashioning an independent operation that serves several different types of institutions, meeting their diverse needs and bringing benefits to all. Though there are problems, there are also many reasons to be encouraged.

If you consider a shared resource center as a business operation, it is clear that you need to *define the products and services* that you are going to sell. You have to understand something about the *characteristics of the market;* you have to identify those specific services that would have general interest and, therefore marketability. You also have to *assess in a realistic fashion both the intent of prospective customers, clients, or users and their capability to buy.* It is not only what the particular user might want but what his institution might be able and willing to pay for.

While on this subject of definition of products and services, a digression is in order to say a few words about the Science Center and how it relates to this particular institutional market. The Science Center is a not-for-profit corporation created in the mid-sixties in the Delaware Valley with headquarters in Philadelphia. The shareholders are twenty-six universities, colleges, and medical schools,* all of which are represented on the board of directors.

* The shareholding institutions are as follows: Bryn Mawr College; The Children's Hospital of Philadelphia; Delaware State College; Drexel University; Hahnemann Medical College and Hospital of Philadelphia; Haverford College; Lafayette College; Lehigh University; Lincoln University; Medical College of Pennsylvania; Mercy Catholic Medical Center of Southeastern Pennsylvania, Misericordia Division; Mercy-Douglass Hospital; Pennsylvania College of Optometry; Pennsylvania Hospital; Phila-

The Science Center is engaged in a variety of activities. We are developing twenty acres of urban land under a contract with the Philadelphia Redevelopment Authority. We are building our sixth building. Just finished, and now being occupied, is an eighteen-story gateway tower. More than 2,000 people now work at the center for its fifty resident companies and organizations. When the $200 million cluster of facilities is completed about 15,000 people will be employed. We are now planning a major residential conference facility, which we hope to have under construction during the course of this year and which will provide a new home base for UNI-COLL. The Science Center as a corporation also plans and manages applied, action-oriented research that uses faculty from many institutions in the area and also cooperates with industry. We serve as an interfacing organization. We are a stand-alone organization.

One of the earliest objectives of the creators of Science Center concept back in the mid-sixties was to provide mechanisms whereby institutions could share costly facilities and not only improve the quality of services but also somehow reduce their costs. Thus, the creation of UNI-COLL as a wholly owned computer subsidiary represents the achievement of one of the original objectives.

In attempting to define a market for specific services, we found that the market was real but was very soft. It was real as a result of rapid growth in the sixties of the use of computing in instruction, research, and institutional administration. The National Science Foundation deserves much credit for its support, which encouraged greater usage of computers in various ways and encouraged the development of small experimental networks. It also encouraged the generation of data bases and of information retrieval systems.

This kind of subsidy rapidly evaporated at the end of the sixties, however, and there was an urgent need to seek alternatives for funding. It was this search for an alternative that led to intensified study among Science Center shareholder institutions of whether this was the right time to move forward toward collaborative and shared use of computing resources in the area. Over an eighteen-month period we held a number of meetings with faculty members, and institutional staff throughout the region who were potential participants and users of a shared com-

---

delphia College of Osteopathic Medicine; Philadelphia College of Pharmacy and Science; Philadelphia College of Textiles and Science; Presbyterian-University of Pennsylvania Medical Center; Swarthmore College; Temple University; Thomas Jefferson University; University of Delaware; University of Pennsylvania; Villanova University; West Philadelphia Corporation; Widener College.

puter resource. It was clear that there was considerable interest in various types of services a regional center could provide. The market appeared to be real; but it was not firm and dependable. It was also difficult to find where the locus of decisions about commitments would lie within a given institution, and that could change from time to time.

The decision was reached, however, to proceed and to create UNI-COLL as a wholly owned not-for-profit subsidiary of the Science Center. This decision was based upon commitments from presidents of the University of Pennsylvania, Drexel, Villanova, and Bryn Mawr, along with the clear intent on the part of other institutions to plan with UNI-COLL for shared use of its services. These presidential decisions had followed many planning sessions involving computer center directors and faculty users on these campuses. Once the decision to proceed was made, it became much easier to focus sharply on the nature of computing services to be provided and the extent of the market for such services.

The second point listed for consideration of a shared resource as a business is the creation of an appropriate business organization. Considerable discussion centered on the question of the structure and the form of UNI-COLL as a new cooperative and shared utility. Very early the private-for-profit route was considered but instead we chose to be organized as not-for-profit. Some suggested that we follow the example of the Triangle Universities Computing Center in North Carolina. But our situation was quite different. In the Delaware Valley we had small liberal arts colleges, state colleges, large publicly supported universities, and private universities. Some pressed for the board of directors of the corporation to be composed of the directors of the computing centers on the campuses. But it was clear that no one person could speak for computing on a given campus. Out of meetings with the users, which also included computing center directors, we attempted to define the functions they all should play in helping to guide developments of UNI-COLL. Their roles were seen to be critically important, particularly regarding operating policies and services.

It was decided that the board of directors of UNI-COLL would be composed mainly of presidents of participating institutions. But then the question arose, how do we get the management set up? Do we take over a management team from an existing university computing center and use it entirely, or do we import a new staff? We gave consideration to entering into a management contract with a private for-profit company in the field of computer service. The grapevine served to carry our interest throughout the country, and we received fifteen proposals, eight of them substantial. A group made up of board members, faculty members, and computer center directors reviewed all proposals. We concluded that all such arrangements would have required UNI-COLL to give up too much control. It was going to be difficult enough at best to provide the kind of responsive service needed to satisfy academic needs without adding problems of another kind.

Thus, the UNI-COLL board decided to move forward independently, to stand alone. It is in this area of being responsive to users' needs that UNI-COLL differs in its operations from a for-profit business that creates a product to be sold in the open market. The institutions participating must be represented not only by presidents on the board but by some of the faculty and staff using computing services. We created an Operating Policy Advisory Committee (OPAC) including directors of computing activities on campuses and faculty and staff users. OPAC provides a forum for discussion of common goals, common program packages, hardware changes, and systems changes. Its members are still refining their mode of operation. OPAC will continue to be an evolving and changing mechanism to provide effective two-way communication between participating users and management. The members and others form subgroups of users concerned with common interests such as interactive languages, administrative programs, number-crunching applications, and student records and registration.

Administratively, UNI-COLL has the usual business type of line and staff organization. I serve as the president. The executive vice-president, Robert L. Logan, is in fact the executive officer in direct control of all operations. Reporting to him are the director of technology, the controller, the manager of marketing, the general manager of programming systems, the manager of applied technical support, and the director of operations. The authority for administrative decisions rests with the corporation, but those decisions reflect continual input from the users and they implement policy that is set by the board.

The most critical problem we have had to face is initial capitalization for meeting start-up costs. When we were exploring a joint venture with commercial groups under a management contract, one of our requirements was for them to provide the initial working capital. In return they wanted control. This was to be expected, but we could not give them the degree of control they wanted because we were moving into an operation that had not been fully tested and we needed flexibility to enable us to respond effectively to needs of users.

We are gaining our working capital through a line of credit at local banks, secured by collateral guarantees from user institutions. It would be far preferable to have been able to secure a grant for this purpose in lieu of using debt financing and thus eliminate the financial burden of debt service. In this respect a not-for-profit organization has a severe handicap in operating in a businesslike mode. It cannot secure equity financing and must rely on grants or on borrowed funds for working capital.

The University of Pennsylvania deserves major credit for enabling us to move forward. It did so by making very hard decisions at the highest level. It was realized that with the expiration of the National Science Foundation's support grant to the university for its computing activities an alternative approach had to be found. The need for computing was increasing in scope and in power, and costs were increasing. Nevertheless, the university decided to work with and through the Science Center to develop a shared regional computing resource. Penn's large commitment for usage, together with those from other institutions, enabled UNI-COLL to go forward.

I would like to emphasize here that the potential for growth of usage by several institutions rests on the fact that UNI-COLL serves them as a neutral and central organization. This circumstance provides a far greater incentive than would management and control within one institution. I believe this aspect of neutrality is one of the most valuable intangible assets of the Science Center in many of our activities. The Science Center serves, for example, as the grantee for the Greater Delaware Valley Regional Medical Program, involving 6 medical schools and 206 hospitals. Those of you who are acquainted with interinstitutional competition can appreciate the difficulties as well as the importance of achieving collaboration in such a group.

Other institutions in the area, in addition to the University of Pennsylvania, realized that they had similar problems relative to computing. Bryn Mawr, Swarthmore, and Haverford had worked together in a limited computing consortium. These three institutions purchased a 360/44 computer system with the aid of NSF funds. It became obvious that the system could not provide what they needed in the future, and yet it strained their budgets. Bryn Mawr and Swarthmore now look to UNI-COLL for computing services.

The fourth element I listed earlier as important in managing a shared computer resource center as a business related to promotion and marketing. Much that I have already said deals with this important function, but it merits specific attention.

No two institutions are alike. Even on campuses having sophisticated users there are many nonusers who could productively and creatively use computing in their teaching, research, or administration. Many small colleges have been exposed to computing only in very limited ways.

Furthermore, we have to recognize in promoting and marketing that there are few captive markets, unless the top administration of an institution dictates that money will not be spent in any way on computing except through a particular center, in this case the shared resource center. In the start-up phases of creating a shared computer resource center such a commitment by top administration is critically important, and steps should be taken to secure it.

Alternatively, some have suggested that institutions give users money to shop around with. In a stable regional situation with a well-established shared resource center this policy might work. But without an institutional commitment for usage the market remains unreliable and taxes the resources of those developing a shared resource center.

For example, at one institution the president, the provost, and a committee of faculty, including the director of the computing center, three times agreed to install a terminal as a start in using UNI-COLL's services. Three times that terminal order was canceled. That institution will, we expect, ultimately become a major user, but time has been required to clarify how and in what way a decision to use a shared regional center can be made to stick. One of the fundamental factors was the extreme difficulty of making credible comparisons of costs and quality of service, because traditionally institutions have not maintained full cost and performance evaluations. The inevitable delays and uncertainties that result from such continued marketing efforts are factors to be recognized in dealing with this market.

As another illustration, another institution made a commitment to spend funds at an agreed-on monthly rate for computing services. But the usage of UNI-COLL there has not approached that level, even though the users are apparently satisfied. We are still trying to find out where institutional diversions are occurring in the pipeline for funds.

Thus, the task of promoting and marketing in the college and university environment involves some special problems for a shared resource center. We are making progress. This month we are scheduling a meeting at which UNI-COLL will be making a progress report on six months of independent operation, including plans for the future. At that same meeting will be presidents, vice-presidents for finance, vice-presidents and provosts for academic affairs, and directors of computing activities from several user institutions. The purpose is to present UNI-COLL's story in a coherent and comprehensive fashion to all such persons at the same time. We believe this will be an important meeting to achieve commonality of thinking and understanding. We expect it to generate further institutional usage and commitment based on proved performance and prospects for further economies and expansion of services.

UNI-COLL now serves 200 organizations, including 15 universities and colleges, representing widely varying levels of usage. This number will grow, as will usage by each institution.

In principle, a shared computer resource center should rate very highly as measured by quality and variety in the production and delivery of products and services. The services should be available at attractive prices, and they should relate to users' needs and be reliable. We are making good on our pledge to do just that.

The production function includes, of course, hardware, software, and people. Last summer we installed a 370/165 system with 2 million characters of processor storage and with 2 billion characters of disc storage. This present system replaces the 360/75 system we inherited from the University of Pennsylvania at the time Penn made the commitment to move into use of a shared resource. We have seven high-speed tape drives, three high-speed printers, and plotting equipment. We serve now 22 remote-batch terminals and 100 keyboard terminals. Of total production, 75 are for service to remote terminals: high-speed, low-speed, and interactive. A full library of systems documentation is available for users. We conduct workshops and courses of instruction to help users and potential users get acquainted with how to get on the system and how best to use it. We are seeking to achieve cost-effective management, as any business would in our internal operations and also to help users make optimal use of the systems.

In all aspects of our business approach to operations we have set management objectives. A major objective is to achieve financial stability. Only in that way can we remain independent and have the flexibility to lead toward new and improved hardware and software systems to bring even more economies and service benefits to our users. Another objective is continual improvement in the quality and reliability of all services.

Furthermore, we must be constantly responsive to changes in the marketplace. Deep and pervasive changes are in progress throughout higher education. They range from new approaches to fiscal and institutional management to changes in the requirements for completion of work for degrees. The nonstructured degree, open universities, multimedia learning systems, credit by examination, and computer-aided learning are but a few signs of changes in the making. We do not generate such changes, but UNI-COLL must be prepared to respond in a responsible way to calls for assistance in such areas.

Finally, as all organizations must, we must work everlastingly hard at improving internal and external communications. All staff must be trained to hold service to users as the prime goal of their thinking and action. They must have a broad comprehension of the policies and capabilities of UNI-COLL. Of equal importance, we will continue to use many means to communicate with present and prospective participating users. Technical bulletins, printed announcements, extensive applications documentation, individual visits and consultations, special meetings with groups, OPAC, seminars, workshops, training courses—all are now being used, and their use will increase.

Promotion and Economics of a Computing Network—the Role of a Shared Computer Resource Center

Although so far my remarks have related to a shared resource center, I believe they have direct relevance to a computing network because they have special roles to play in a network. These roles are to provide a stable system for management, to promote expansion of use of services, to serve as a base for stimulating interinstitutional collaboration, and to provide financial control points.

Shared resource centers would provide for management of special programs that could be made available throughout the country via the telecommunications and other interconnecting systems of the network. This would include management of special data bases and maintenance of ready regional access to the network.

Second, the regional centers would provide a number of services geographically close to users. Some of the other papers refer to the importance of users having ready access to systems programmers, for example. Although in principle a user at Harvard might communicate with a systems programmer at UCLA, it would be rather costly and impersonal. The three-hour time zone differential would also present problems. There is thus genuine merit in having geographic and regional proximity for access to consultation, documentation, and trouble-shooting in general. The shared computer resource center can be a major force for promotion of use, for an expansion of services—to help create a market, if you like. It can also prepare and staff educational activities within the nearby small and large institutions. Small institutions particularly need this kind of personal attention.

Third, the shared resource center can be of great assistance to a national network by generating habits of resource sharing inside and among institutions. The behavioral problems inhibiting resource sharing are really much more severe than the technological ones. Institutional habits and commitments to the concept of sharing must permeate all strata of the institution, top administration, schools, departments, and individual faculty and staff as users. Through such regional collaboration special data bases might also be generated and made available through the networks. Furthermore, new applications programs developed by common action of innovative faculty in a region can be proved out and then be made available for network use.

Another role, and in my judgment a very important one, is for a regional shared resource center to serve as a control point for assessment of charges and billing and for accumulation of the cost data essential for evaluation of performance. We have all been frustrated in planning shared facilities, and now in planning networks, because of the lack of reliable cost and performance data. Fundamental to my proposition that all shared resource centers be operated as a business is the need to generate realistic financial data. Comparisons among alternatives leading to top-level decisions require such information. Such regional centers will be concerned and responsible not only for proper accounting relative to their regional users but also for charges incurred through network use. These data can be used in continual review and evaluation and can lead to improvement of performance and extension of economies to users.

The problems and the challenges we have encountered in creating UNI-COLL either have been or will be experienced elsewhere. No one appears to believe that there are simple answers to the problems, the most severe of which are behavioral in character. But I am convinced that we can make progress toward our goals, particularly if we bear in mind that we will have to face up realistically to certain issues.

Management of Shared Computer Resources

The following are brief statements of four such considerations:

• *New roles for institutional directors of computing activities will emerge.* No longer will they manage large hardware and software systems with associated personnel. Instead they will have a professionally challenging task of improving use of computing throughout all sectors of the campus, taking full advantage of a much richer resource of services through the external shared computer utility.

• *Advantages through economies of scale and a wider range of applications programs and services through the shared resource will have to be accepted as beneficial gains offsetting loss of having all systems located on each campus.* Intelligent and powerful terminals as well as a diverse set of simple low-speed terminals are becoming, in any case, the principal interfaces between users and powerful computing systems wherever they may be located.

• *Traditional institutional approaches to budgeting and cost evaluation of computing services will have to give way to treating computing as a line item operational expense in virtually all operating divisions of an institution.* This becomes a controllable and predictable budget item when treated as an expense, covering charges incurred through use of an external shared computer utility. No longer will an institution face the uncertainties of capital costs, deficits in computer center operations, and an uncontrollable impact on overhead.

• *Success of a regional shared facility will necessitate recognition that it is inherently a shared responsibility as well.* This means institutional commitments to share through usage. It means collaborative planning and action. Also it will result in delegation to the neutral regional organization of the responsibility for management, along with the authority to make administrative decisions. These decisions, in turn, should reflect good business practices and policies and the continuing flow of recommendations from all users.

I feel confident that we are all moving in the right direction, at the right time, and with the right instincts as to what must be done to bring benefits of computing throughout substantially all instituions of higher education.

37

Computers and Communications

Report of Workshop 9

JULIUS ARONOFSKY
Southern Methodist University
Faculty Discussion Leader

Contributors:

WILLIAM BOSSERT
Harvard University

IRA COTTON
National Bureau of Standards

DAVID FARBER
University of California, Irvine

WILLIAM HUGGINS
The Johns Hopkins University

LEON KATZ
University of Saskatchewan

DAVIS B. MC CARN
National Library of Medicine

CLAIR G. MAPLE
Iowa State University

STUART L. MATHISON
Bolt Beranek and Newman Inc.

JOSEPH B. REID
Université du Québec

ROBERT ROBERSON
SUNY at Binghamton

Introduction

Workshop 9 considered alternative network designs, computer equipment configurations, telecommunications approaches, switching systems, and basic network configurations. Special attention was focused on the capital cost of alternative approaches and their implications for operating methods, costs, and the ability of the network to serve important users' needs. In addition to evaluating present and prospective hardware configurations, the workshop dealt with questions of telecommunications policy, since it seemed likely that policy will have as much impact on the economics of alternative network approaches as will the technical state of the art.

The participants felt they should not go into hardware design in any depth, since existing computer and communications technologies were more than adequate to initiate plans for a national science network. Because existing technology may be quite obsolete by 1980, this report deals with short-range utilization of the existing technology while noting the technological innovations expected after 1980.

In the shorter range, the workshop suggested that with one exception to be discussed no significant new investment was needed in communication technology for the purposes of a national science network. Networks do exist in a variety of hardware technologies, and we ought to take advantage of them, especially since no one was aware of any new concept that offered a clear enough advantage to justify the cost of its development for the network.

The workshop favored design criteria that would reinforce the network's being heterogeneous, acentric, and general purpose:

Heterogeneous, because only a network of many makes of computers can provide a marketplace for all computer services and exert effective pressure toward standardization. This should not lead to chaos, as some might fear, but to increased stability.

Acentric (without a central computer), because only an acentric network is capable of indefinite expansion involving many institutions of diverse interests.

General purpose, so that resources could be shared among many functions and serve many communities of interest. We should not regard a national science computer network as our target but rather an international computer network.

Many existing computer networks do not meet these three criteria. Many are not acentric and many are special purpose. The workshop felt that existing networks must be used as building blocks rather than prototypes of a national network, and the cement for these building blocks should be the ARPA Network technology, with proper modifications. The ARPA Network in its present realization appears expensive for attaching small computers or terminals, but not for interconnecting the large computers that form the centers of the many star networks. This was the chief reason for the workshop's recommendation.

Specific Proposals

The workshop offered a number of proposals that it believed would help provide a framework for advancing national educational networks. It recommended:

1. **That several of the existing regional networks, such as TUCC, UNI-COLL, and MERIT, be linked together to take advantage of the ARPA technology.** Many of the twenty-five star networks funded by NSF should also be involved.

2. **That a general-purpose network such as ARPANET be linked to one or more of the commercial computer utility networks.** ARPANET and/or some of the regional networks could be linked to commercial networks through a common host computer. No doubt a number of important technical and operational problems would have to be solved in making such interconnections.

3. **That a local network of small processors and minicomputers become part of the overall design.** The Distributed Computing System (DCS) with its telecommunications ring, which is under development at the University of California, Irvine, is an example of such a local network. The key concern here is to make sure that the small user does not get lost in the shuffle and that he is able to use the proposed network.

Developments

A subgroup of the workshop was asked to comment on ways to extend the utilization of existing networks to suit small-institution users and to provide estimates of development costs, if possible.

The subgroup's objectives were (1) to explore the potential for connecting existing networks to the ARPANET in ways that would be least likely to upset its operation, and (2) to extend network access to environments not now economically feasible.

To do this, the group proposed to make use of proved techniques and existing hardware, even at the expense of some inelegance of implementation. Even so, some unresolved needs will have to be met and some innovation will be required in two of the proposed configurations.

It seems desirable to explore interconnection possibilities at several different levels—namely, in order of decreasing priority, those of the regional service center, the local network of small processors, and the non-ARPA network of computers. The computer facility of the small educational institution should also be included. Each of these projects is discussed briefly now.

Regional Service Center
At present a substantial number of cooperative, regional computer centers are operating with varying degrees of success. An important question is how such centers would make use of, and be affected by, access to a national network.

These service centers could serve as natural entry points to tie their already developed constituency to a larger network. In principle, such access would enable the service center to handle peak loads that would otherwise overload it and to broaden the available computational services by making use of the network's resources.

A dominant objective would be to accomplish this aim in a way that would not upset the existing operation and reliable functioning of the service center. Even if technical problems do not arise, managerial issues will have to be faced in incorporating the network services into the operation of the center, and experience with how this is to be done will be valuable in planning further developments relating to a national science computing network.

Several regional service centers could provide appropriate connection to the ARPANET through an IMP either at a local or at a very remote site. Network control programs already exist for this purpose, and hence this tie-in would require virtually no technical development. The cost of this project would be roughly $100,000, including the purchase of the IMP.

Local Network of Small Processors
Providing access for an existing local network of minicomputers (for example, DCS) into a larger, general-purpose network could provide an interesting setting for the investigation of issues relating to hierarchical computing. The minicomputers could perform data acquisition and limited local processing, calling on the general network for specialized computing resources (for example, large program packages or data bases) are required.

The group believed that access for a network of minicomputers would most easily be provided by the development of a special network interface processor. This processor would serve as another processor in the minicomputer network. All requests for outside network access would be directed to it. This computer would also probably contain the network control program for connecting to the ARPANET via a very distant host interface.

If a minicomputer is chosen for which an NCP has already been developed (such as the PDP-11), we estimate that this project could be completed with a funding of about $50,000.

Non-ARPA General-Purpose Network of Computers

Interconnecting two general-purpose networks of computers will force a resolution of the protocol incompatibilities that have arisen between different networks. Investigation of this problem is essential to the development of any comprehensive and effective nation (or international) network.

Although this resolution would ultimately be best accomplished by use of an appropriate internetwork coupling processor of the same general sort as recommended for the minicomputer network, it would require more extensive (and riskier) development than seems necessary at present. Instead, the group recommended interconnecting the two networks through a host system common to them both.

Dissemination of Information

Workshop participants were concerned with the need for information facilities that relate to hardware availability and usage. They adopted the term "user and server brokerage information" to mean a service dynamically available in as automated form as possible. It would include complete catalogs of quality, price, and status of procedures and processors. To a limited, inadequate extent this capability is available on the ARPA Network.

The service should

1. Act as a broker between the user and provider of services and allow the receiver and provider on-line to match the need to available resources via access to the network. Information on available facilities, prices, service quality, hardware conventions, and the like should be dynamically updated so that a user or his program could be led to the most appropriate supplier, given fluctuations in demand and failures in this system.

The group believed that a user's service request should yield a response detailing procedures for fulfilling it on an intra- or internetwork basis. The response procedures might include addressing codes and standard formating conventions that the user could feed back to the network to get the appropriate access, with data conversions as necessary by network interface, without any detailed understanding on his part.

2. Include a brokerage service between networks similar to the one it provides between user and server. The interface between networks is, in fact, the only logical place for the programs and data base comprising the broker information function. Presumably the information on each terminal and host computer within a single network would be current on the facilities and procedures within the net. Therefore, the only need for information would be for users passing from one network to another.

3. Provide information, in terms of network requirements, on gaining access to any given network or on crossing networks to obtain services.

4. Aid in the establishment of standards based on the reaction to existing requirements for reaching the network.

5. Interact between users and servers in determining the level of users. This documentation might consist of codes to provide automatic protocols from one network to another, or it could be extended to teaching materials at a general level in a variety of media.

At least some research is imperative before conclusions can be reached on how some of these functions could be realized. The following list indicates the types of questions that must be evaluated:

1. What level of documentation, either in terms of applications or network requirements, is required for successful usage?

2. Where should the responsibilities reside for generating documentation —with the creators of the application or network, or with the broker or information group that will make the information available?

3. What form of on-line information system should be designed to allow users to acquire information compendia such as catalogs or dictionaries?

Although automated information retrieval capabilities are needed to facilitate internetwork communications, the workshop did not feel that every conceivable user request should be satisfied with an efficient, idiotproof system. The users must rely on clear statements of the limitations of the retrieval system for internetwork conversions. The responses must be consistent with the current system status and the sophistication desired by the user.

Telecommunication Policy

Existing networks are private in nature and present telephone company tariffs and FCC communications policy prohibit their use for furnishing public communication services. The Communications Act of 1934 provides that if these networks are used to sell communication services outside the operator's own organization, the secondary organizations must operate as regulated common carriers. They must apply to the FCC for construction permits (if their service is interstate or international) and, if authorized, file tariff schedules with the FCC describing the price structure and the conditions of service.

There are two basic exceptions to this requirement. The first occurs where the organization is providing remote access computing or information services for hire and uses communications facilities leased from an authorized common carrier as part of the package service offering (for example, a commercial time-sharing service firm). This is analogous to a department store that operates a private trucking fleet for deliveries but does not sell trucking services as such to the public.

The second exception is exemplified by Aeronautical Radio, Inc., which provides communication services to the United States domestic airlines. The FCC has allowed it to operate as a private activity rather than as a regulated common carrier. The justification, in this case, was that the ARINC system was highly specialized, served a restricted user community, was operated by a nonprofit corporation, and, most important, was necessary to assure safety of life and property.

Thus the present status of communication law and policy provides for three alternatives for a national science computer network:

1. One or more total package services providing communication services as well as computing and information services. Each package service would be provided on an unregulated basis.
2. A collection of independently provided computing and information services with remote access through an ARINC-type communications subnet. The subnet would be unregulated, as would the computer services.
3. A collection of independently provided computing and information services with remote access through a regulated common carrier communications subnet.

The first and third alternatives are clearly established in the body of United States communications law. The second alternative is not now well established as a precedent, since the ARINC system is unique. Thus, it may be difficult to justify a choice of this alternative. The issue is certainly not settled, however, and deserves further consideration and study. If such an investigation is undertaken, the limitations of an ARINC type of activity should be carefully considered. These include its likely restriction to educational and research institutions, excluding industrial and government servers and users.

The most practical alternative appears to be the establishment of a suitable common carrier organization. In that instance a number of practical questions arise. Would it be desirable, for example, to establish one or more of these specialized networks? How long would the FCC take to decide? How should services be priced? Is it possible to delimit the user community? Can backbone transmission facilities be provided by the established telephone carriers? Other questions will surely emerge as these issues are explored. Given the probable common carrier requirement, it appears timely to explore these questions.

Beyond the 1980–1985 Period

It was not the workshop's intent to devote a great deal of time to technological innovations that are expected beyond the 1980–1985 period. We felt it more important to ask how to get started now with existing technology. Nevertheless, we would have been remiss if we had not explored future directions and opportunities that will render some of the existing technologies obsolete.

The relationship of any present computer communication network to future systems cannot now be predicted. The development of university computer networks will probably not be a dominant force as networks find increasing commercial and industrial uses. Alternative futures depend heavily on cost trends.

In *The Economics of Computers* William Sharpe pointed out that the speed of central processors has been increasing by a factor of 10 every five years while the cost per machine cycle has been decreasing at the same rate.[1] This change, coupled with full digital-communication long-distance lines now projected by the common carriers, satellites, and LSI technology-megamemories, will have a profound impact on computer networks. An even more profound impact would result from the introduction of two-way broadband communication into the home, providing on-demand entertainment programs in color and with a stereophonic sound. The growth of such systems would tap an enormous source of wealth in our country, the consumer, and give rise to a national broadband network reaching into every home.

Once such a system was established, it would probably overshadow all others. All other network activities, including those for education, commerce, and health, would be readily handled by this massive system.

But the rate of decrease in computer costs versus the rate of decrease in communication costs will profoundly affect the trade-off between mini-computers and megacomputers. The outcome of these trends cannot now be predicted. Moreover, the Bell System as well as the emerging special-ized carriers will probably provide switched data communication services whose impact cannot now be foreseen.

Satellite Communication

The domestic communication satellites offer a potential to interconnect regional computer networks at low cost. Such interconnections could be established regardless of the distance between the networks. Satellite technology has advanced to a stage where the cost of the equipment (transmitters, antennas, and modems) may be as low as the present net-work connection cost. In April 1974 the National Aeronautics and Space Administration will launch Applications Technology Satellite-F (ATS-F), which will have a broadband relay capability. If practical, the NSF should move quickly to develop experiments for using this satellite to connect computer networks. The results would be important in advancing the technology and formulating policy on both domestic and international satellites.

CATV

Since the resolution of the remote signal rebroadcast issue, cable tele-vision systems are expanding to such an extent that the United States is moving toward the status of a wired nation. This expansion promises to introduce broadband communications into the majority of American homes. The cost of processors and terminals is dropping rapidly. Thus, the potential user group for network services could be expanding rapidly to provide services to home and office users (for example, the TICCIT system of the MITRE Corporation). Projects should be initiated to explore the use of CATV as the local communications media for com-puter networks. The need for such home service must be examined, however, and experience developed for assisting policy formulation and providing guidance for the negotiation of community franchises.

Reference

1. William W. Sharpe, *The Economics of Computers,* Columbia University Press, New York, 1969.

38

Software Systems and Operating Procedures

Report of Workshop 10

JAMES L. MC KENNEY
Harvard Graduate School
of Business Administration
Faculty Discussion Leader

Contributors:

WILLIAM ATCHISON
University of Maryland

STOUGHTON BELL
University of New Mexico

HOWARD CAMPAIGNE
Slippery Rock College

KAREN A. DUNCAN
Medical University of
South Carolina

ROBERT R. KORFHAGE
Southern Methodist University

E. REX KRUEGER
University of Colorado

PHILIP L. LONG
Ohio College Library Center

GREGORY MARKS
University of Michigan

JAMES POAGE
Princeton University

WILLIAM WALDEN
Washington State University

GLEN J. WIEBE
Associated Colleges of
Central Kansas

Introduction

This workshop considered software systems and operating procedures
as a key means to providing service to the user. It did not look into the
intricacies of relating one computer to another or a computer system
to a particular network. More important, it considered how users should
operate in relation to a network and what are the trade-offs for the user
of consistency versus effort.

In the process the discussion considered a series of design issues affecting
any software/hardware system. It was assumed that to develop service-
oriented networks will require significant developmental costs and that
to offset them the system must appeal to a large number of users. In
part this discussion was influenced by experiences including difficulties
in existing network systems, particularly the different requirements for
different machines on the ARPA Network, the wide variation among job
control language systems today, and the futility of creating one more
planning committee on standards.

General Operational Characteristics of Network Systems

The workshop agreed that the network software system should create
the impression that the user is controlling his own computer. This
assumes that the user will work through some form of terminal or
remote job-entry instrument connected to a computer or a communication device. The device can inform the user of the status of his job and
the impact and implications of any failures for his job. The network
itself should be transparent and should not intrude on the individual's
operation in getting his computing service as he desires. It was suggested
that part of this goal should be to make the individual's operating
procedures identical, no matter what computing source he relies on.
There might be, as an example, a standard sign-on procedure for the
network independent of any computer. The process would also be such
that unless the individual specified a particular machine, the standard
protocol and computer network would find the identifiable source similar
to Farber's ring computing.[1]

An important ingredient of this transparency must be a general-purpose
accounting procedure that fairly and consistently accumulates the
charges of any particular user no matter how and where he incurs them
and that records these charges so he can understand them. Another requirement is a standard library of user files that allows the user to determine the location and content of his files and a file inquiry system that
is as convenient to use as his own system would be. To support en-

lightened usage, there should be an understandable indication of the costs of service for a given period of time. The cost factors could be updated periodically, but they should be accessible and understandable to the user.

There is a strong feeling that standards should exist for files, calling sequences, linking, and the like, but there is frustration and concern over how one implements such standards. The best tack seems to involve informal cooperative ventures that identify a set of rules various people can observe, implement those rules on a network, and try them until they are so common that they are accepted as standards. In general, the workshop felt that this strategy could work, particularly if the National Science Foundation supports it.

In discussing the opportunities for developing such a user network, the workshop set a few ground rules for its deliberations. The first was not to attempt a rigorous definition of system specifications for actual implementation. The time was too short and the knowledge too thin. Rather the aim was (1) to develop a descriptive set of concepts that could be useful as aids in the design of operating procedures for a system, (2) to define a few unanswered questions that would suggest ideas for research, and (3) to explore possible software procedures for network users.

Organization of Workshop Subgroups

The workshop developed four parallel topics in its discussion of how best to develop conceptual frameworks to aid in the design. The first and overall guiding principle was the differentiation between logical networks and physical networks. A second specific topic was the specifications of a network control system that could aid the user in entering, finding the requisite service, accounting for his service, linking files, protecting his privacy, and aiding in recovery from either computer mishaps or communication failures. A third difficult software problem is file maintenance and management in a network system. Finally, the functions an automated user interface in the control system could perform was explored.

In brief, the workshop's conclusions were

1. A logical network is the appropriate basis for the planning of networks to service the needs of higher education.
2. The software interface between the user and the network computer must be designed to serve the user and manage the system.
3. Special consideration should be given to file maintenance and security of information in designing software.

4. A network interface device that allows local information and/or system control is an important need.

Discussion of Workshop Summary

The first step in a design phase of expanding network usage is to articulate systems as logical networks, relating them coherently and consistently to a set of tasks. A logical network consists of a set of programs and data bases related to a particular coordinated group of tasks, for example research on crystallography or musicology. It may be part of a physical network of computers, control functions, and terminal lines; a complete computer network; or part of several networks. The implementation of the logical network would be through a control system that would accept inputs from the user and distribute them to the appropriate computing resources while accounting for the user's charges and maintaining records of the user's files. It appears that the definition of an appropriate set of logical networks for higher education is the first step in comprehensive planning for the establishment of the sharing of resources. A planning effort could then proceed to determine how such a system of networks could be implemented.

The control network responds to the user and functions as manager of the total system. This control network should provide a standard format and should have a broad range of default conditions the user could invoke to get a variety of service if he is informed and eligible. The management system might simultaneously deal with the user and monitor his activities to collect information for improving its service. A vital aspect of the network would be the documentation, protocols of use, and human support well identified and competently trained.

The workshop group exploring the concept of a logical network felt that the idea was an important base for the design of network control systems, since it emphasized common needs that, if met, would reduce the overall complexity of security, file maintenance, linking, and other activities. The thesis is that looking at users with like tasks allows certain advantages in software design to be implemented by taking into consideration the common set of information-processing activities. The particular flavor of this assumption should be investigated, for it probably includes subtle long-range implications.

A logical network undeniably presents a vital issue of implementation. Communicating in some consistent fashion with the user as well as translating his inputs to a variety of computing and communication devices is a very complex problem of software development. In essence, it is a translation process that is hardware with some software variation. In the short run we need research into the limits of user needs and how best to meet them. In any event, it is going to cost real money to develop and implement.

The Network Control System

Star networks and resource-sharing networks have been in existence for some time, and some of their properties are understood. An end user of a star network visualizes his terminal as an extension of the central computing site. All the documentation, systems, and protocols associated with the central site are imposed on the user. The absence of flashing lights and spinning tapes is of little concern. For the most part, a user feels he has a general-purpose facility at his fingertips.

As use of information processing has spread into a variety of disciplines, problem-oriented or discipline-oriented systems have evolved that give the user the feeling that he has a special-purpose facility at his disposal. Special-purpose systems tend to develop maintenance problems, but more natural communication with the user more than offsets the cost.

Resource-sharing networks are now visualized by users as a collection of star networks that may be entered through a common communications system. As a result, the user must have knowledge not only of the network documentation and protocols but also of the documentation, systems, and protocols of every resource he may use on the network. The ultimate goal in the evolution of complex resource-sharing systems is to provide the user the appearance that the network is one resource described through one set of documentation and subject to one set of protocols. For task-oriented users, the network must function in the natural language without losing the power inherent in a distributed network.

To give a network the appearance of a one-to-one relationship with a user as a contemporary time-sharing terminal does, the group postulated the necessity of a control agent. This agent would provide a uniform control language and uniform pricing via hardware, possibly of special design, dedicated to controlling the user's interaction with various sources of computation power and information.

The control system would accept input from a variety of terminals and translate the control language into statements appropriate for submission to the network's service processors. It should also permit the user to supply his own "native mode" control statements, as he likes after he has given the standard sign-on and has identified the interactive target processor or processor class. The control system might be commanded to select an appropriate processor, given run constraints and machine-independent job descriptors. The control would also permit the user to inquire about the status of his job or cancel its execution at any time.

The user could sign off after his run was submitted to the network and log on later to accept his output and pertinent accounting data. During a terminal session the user could consult extensive system documentation via the terminal or could simply seek brief prompting messages as the need arises. The network could also link the user to a consultant if necessary. Finally, a system should include emergency procedures that would keep the user informed when occasional crashes or failures hit the communication system. It is particularly important to maintain contact with the user in time of travail.

The software challenge is to build a network control system that will not only allow the user to be independent of particular hardware but will contain language elements that will make it feasible to manage the net during the operation of the system. By manage we mean make decisions on effective use of computing resources available, inform the user of these trade-offs, and improve the management process. Thus, the system would not only involve the normal acquisition of statistics but would also use the statistics to manage the location of files, identification of users and their needs, and the maintenance of an information system on the network's services. This information aspect of the control unit should be parallel with the system for communicating with the individual user. In terms of systems we are suggesting that the network control system is similar to Wyatt's translating system, providing an interface between the user and the computer system. We are further suggesting that such an interface have within it the capacity to make economic trade-offs, inform the user of potential economic trade-offs, account for his services, and manage files.

When a general user approaches a computer system, he needs to gain timely access to it with an adequate amount of control to accomplish a task. A critical element in accomplishing a task through a computer seems to be a certain amount of tender loving care in addition to the software. There are times when documents are not understandable or when problems arise, and the human element seems vital for adequate service. An adequate amount of resources and attention must be devoted to making people available to help people use the system.

A particular problem of network management will be that related to files—the issue of privacy, file maintenance, and beyond that file recovery. A new problem may well be library accountability of files so that one can understand what files are accessible and updated in a uniform fashion for a variety of users. It is particularly important that the user population understand the actual state of the art of file security and maintenance and that efforts be made to develop adequate measures to maintain the integrity of sensitive files.

Much of the technology to implement such a network environment is already available. But serious research is sorely needed on these questions:

1. Can a coherent control system for a network's resources be established for users?

- Machine-independent job description parameters.
- Low-cost network support hardware of special design.
- Access to and interface between different physical networks.
- Network fault detection and isolation.
- Network control language.

2. Is an effective data transmission capability available?

- Machine-independent data descriptors to facilitate data-set use on a variety of machines.
- Data compaction to minimize traffic.

3. What standards are required for hosts to join a network?

- Capacity to serve.
- Security for files, catalog of user files.
- Standard accounting output at end of job.
- Use of common descriptors for job execution specifications.

4. Can the network's monitor-broker be established, and how?

- Human or machine.
- For each network—logical, task, or physical.
- How is sign-on permission granted?
- How is accounting handled?
- Is a review board needed to select the resources to be included?

Given a specification of the operation of the network system, a strong design effort should be undertaken to consider which aspects of the sytem should be automated and put in a "black box" so that when the user signs on the terminal will generate a set of codes that will link files, identify proper formats, and deliver output to the appropriate systems.

An interface management system similar to a terminal interface processor (TIP) but more focused on accounting and information transfer than information communication should be invented. The design effort for this device would include identifying what should be maintained locally, perhaps an accounting system on the user's needs, and what should be maintained centrally. Another question is what should be automated in hardware or maintained in software. It is envisioned that to operate a large expanse of network, local information about user needs, libraries, accounting, and so forth might best be maintained in a small computer. Other information regarding files, account numbers, charges, and the like might best be done centrally. A brief discussion of the trade-offs indicated that this was a real opportunity and a complex problem that warranted further consideration.

Conclusions

The workshop concluded that support should be sought for a planning effort to implement the concept of a logical network for two or three potential user groups now being served partially or not at all by existing computer systems. This effort should be undertaken with the users and systems designers to develop a comprehensive plan for the network. Second, there should be an effort to design a network control system that would service the user and manage the resources available. Such a project exists in part at several of the regional networks. The state of the art of these software systems should be explored as crude control systems. Third, a continuing effort is needed on the problems of file management in networks. Not only privacy but the problems of decentralized maintenance and centralized coalition are important. Fourth, the present architecture of the TIP affords an interesting start to consider a communication device for a network of computers but is not the answer to the user's needs and the needs of a truly transparent network. Some design and experimentation should be implemented that would explore what sort of a network interface would be most desirable for a range of users from specifically discipline-oriented to general-purpose. Fifth, the group concluded that human accessibility to other humans was a vital element of successful network operation in addition to fair and equitable understanding of charges and reliable service. This is particularly true in the development stage of any system the networks are likely to go through in the next three to five years. Even in long-term operating systems, however, it seems to be a vital ingredient. The workshop strongly encourages people to keep this fact in mind when planning future network experiments.

Reference

1. D. J. Farber, "Networks: An Introduction," *Datamation*, Vol. 8, No. 8, April 1972, pp. 36–39.

39

Applications Development and User Services

Report of Workshop 11

MARTIN GREENBERGER
The Johns Hopkins University
Faculty Discussion Leader

Contributors:

EDWARD L. BRADY,
National Bureau of Standards

RUSSELL BURRIS,
University of Minnesota

SYLVIA CHARP,
School District of Philadelphia

JAMES A. DAVIS,
National Opinion Research Center

CRAIG A. DECKER,
The Johns Hopkins University

THEA D. HODGE,
University of Minnesota

HAROLD B. KING,
The Urban Institute

ELIZABETH R. LITTLE,
North Carolina Educational Computing Service

ARTHUR LUEHRMANN,
Dartmouth College

F. A. MATSEN,
University of Texas

JAMES NORTON,
Stanford Research Institute

A. HOOD ROBERTS,
Center for Applied Linguistics

GERARD P. WEEG,
University of Iowa

Synopsis
By "networking" is meant the use of computer-communications technology to facilitate the sharing of information and computer resources over great distances. Networking provides an opportunity to increase communication and intellectual commerce among computer users and to expand the size of the market for local computer products and services. The success of networking depends on these opportunities being realized. Networking also both suggests and requires new incentive structures for the production and maintenance of reliable services and good documentation. The incentives include prestige and royalties, analogous to those received in the publishing of books and articles. There is a need for potential network resources to be identified and cataloged, not only in starting up the network, but in its continual operation and as an ongoing information service to users. Networking makes the adoption of standards for compatibility and quality control more essential than ever. Standards can be buttressed and supplemented by expert refereeing and comprehensive evaluative mechanisms. Networking calls for energetic marketing efforts. The participation and experience of the private sector in this function should be sought and encouraged. Network users are likely to be a highly varied lot, and the widely different requirements they place on the network will strain the imagination of management. Critical issues for policy consideration and management attention include questions of governance, priority, allocation, funding, and law.

Introduction
As was customary in most workshops, the initial period in this one was spent trying to establish ground rules and to agree on terms. This warm-up period exhibited lively vocal expressions of strongly felt views. One subject for debate may have done as well as another for the purpose of release, but it was the concept of a "network" that took center stage. Some wanted to restrict use of the term to the idea of a single national system of electrically interconnected computers. Others wanted to speak more generally of various kinds of activities and arrangements (which was referred to as "networking") wherein separate institutions work together to develop and share resources on a broad scale with the assistance of computer communications technology.

As one participant pointed out, it was a little like arguing about the "shape of the table," a reference to the early stages of the Vietnamese peace negotiations in Paris. Positing the idea of a single physical system may have been a useful way to sharpen issues and focus discussion, but it was not essential for most of the discussion that ensued on the operational problems of networking.

When the participants got down to serious business, they took up such topics as incentives, identification of network resources, standards, quality control, production of user services, marketing, governance, allocation, and funding, first as a group and then by dividing into teams of three. Each team selected two questions on these topics as gleaned from the initial discussions and preparations. The division into teams and the addressing of concrete questions changed the operating style of the workshop markedly. Members of a team continued to interact and work energetically within the team, but no longer so much between teams. Consensuses seemed to develop quickly under the stimulus of the specific tasks that had to be completed within a well-understood time limit. One participant, while acknowledging the necessity for the work orientation, deplored the partitioning and insularity between teams that it induced. He saw a hypothetical parallel in the possible tendency of a national science network to break up into small independent subnets and pointed to the importance of finding ways in networking to preserve group interchange and to develop viewpoints that are broadly based, such as through interactive gaming and Delphi-type conferencing.

Networking
A minimal definition of networking is use of a physical communication system for the sharing of resources among different users. The workshop agreed, however, to go beyond that limited view to mean a coordinated effort to share the production, documentation, validation, marketing, and distribution of computer services and resources among different users. The workshop asked itself how networking could assist in the performance of such functions not as a solution itself but as a vehicle for finding solutions to current difficulties.

There are two classes of difficulty that networking might help to resolve. The first has to do with the lack of communication and intellectual commerce between isolated computer users, the second with inadequate size of local markets for worthwhile computer applications. The lack of communication and commerce results from

• A multiplicity of local idiosyncrasies of language, operating systems, and hardware.
• A consequent lack of easy transferability of programs, documentation, and application manuals.
• The difficulty of locating programs and data bases.

- The difficulty of identifying and collaborating with colleagues at other institutions.
- A lack of adequate mechanisms for review and accreditation.
  The inadequacy of the local market inhibits commerce and results in
- Little incentive to produce, document, or validate programs.
- Little political influence, financial support, or individual prestige.
- Little critical evaluation by specialists at other institutions.
- Little peer group recognition in the local community.

It also means that distribution of a program to a large community of users may require the giving up of ownership and control of the program.

By fostering communication and commerce among a large number of users, a network changes the traditional pattern of computer usage. Networking brings to the fore problems that have needed concentrated attention for some time and offers the possibility for developing solutions. Local idiosyncracies of language and systems are likely to decrease when many users communicate both with one another and with varied computer systems. User groups formed to address problems associated with quality control and evaluation could supply peer group recognition similar to refereeing and publication in a scholarly journal. A widened market for high-quality computer applications could provide a much-needed incentive for improvements in the production and documentation of computer programs and data bases.

A unique opportunity afforded by networks is the ability of authors to maintain ownership and control of a program whose use is made available to a large number of people at remote sites. Significant additional benefits are possible in computer-mediated communication among individuals and groups. The need for improved communication and cooperation is made obvious by the introduction of a network. Without progress in filling this need, networking cannot be successful, and many promising educational programs and research projects of the future cannot be effectively realized.

Incentives

A large network of computer resources and users is like a publishing house. Both are sizable marketing and distributing agencies for the products of individual creation. Both free the ultimate customer from exclusive dependence on the local scribes. Both provide authors with a wide competitive market that stimulates them to produce the best they can. It is not surprising that the workshop, in thinking of incentive structures that would encourage good program development and documentation, was guided by the incentives found in the publishing of books and articles.

First is the incentive of prestige. To the extent that networks use refereeing by professional peers and have a stiff editorial policy enforced by editors of high reputation, "publishing" a program and its supporting documentation via the network is likely to carry a prestige factor not unlike that associated with publication of an article in a scholarly journal. This would be enhanced if the recognition received through editorial accountability was weighed positively in decisions on the granting of tenure to junior faculty members and contributed to the reputation of senior faculty members and research scientists.

Start-up funds would probably be required to establish the initial refereeing and editorial organization of a network, but revenue from network commerce might eventually suffice to support these aspects of the operation. Important evaluation of network resources could be made by the users themselves. A user might conclude a session by answering on-line such questions as: Did the program operate as advertised? Did it provide the expected results? Was the documentation adequate? Statistical summaries of user evaluations could be produced daily and automatically by the operating system of the network for the benefit of users the following day.

A second incentive for authors is monetary. Author royalties for program products have not been simple to arrange in the past because of legal difficulties in applying the conventional protection devices of copyright and patent. A network, however, provides the means for an author or his distributing agent to hold on to a program and to supply access remotely for its execution (but not its copying). The program can in this way be made available to all network customers on a restricted basis, its use metered, and the author rewarded accordingly. The royalty incentive will be strongest once the network is in active use and amply subscribed. Until then, other incentives will be required to stimulate product creation.

Royalties for product documentation should be even easier to arrange. Here the copyright provisions for ordinary printed matter would apply, at least with respect to printed user manuals and supporting documentation. Information made available on-line could be handled like the programs themselves or considered as a value-added feature of the programs. By greatly increasing the number of users interested in a given product, a network could provide powerful new incentives for producing and disseminating the high-quality user documentation that now too often is missing. The private sector could help the author to produce this documentation and make it widely available if the demand and funding were assured.

Another important incentive would be money to help finance the development of new markets and the initiation of new programming projects. Such venture capital could take the form of advances against future royalties, contracts for the writing of documentation, fees for the creation of programs, and grant support. The workshop felt that some percentage (perhaps 5 to 10) of the total commerce over the network should be reinvested in the development of new products and in their documentation, with an even larger percentage reinvested to aid the network's expansion while it is still growing.

Market forces, at least initially, may be inadequate to bring about the high-level detailed documentation essential to proper maintenance and use of network programs. A requirement for documentation to be built into programming languages at the syntax stage might help. If this is done in a hierarchical fashion, users would be able to request information at any of several levels. Methods for producing good expressive documentation should be part of computer courses, and the evaluation of documentation should be a key activity in network operation.

Identifying Resources to Be Shared

There are at least three different needs for identifying resources and informing users of them:

1. The need as part of a careful assessment of the overall feasibility and timing of the start-up of any network, based on the assumption that a substantial amount of resources must be shared to make the undertaking worthwhile.
2. Once the network becomes operational, the continuing need for making additions and removals of resources to and from the network, influenced by funding dynamics, effects of the freemarket, future user needs, user experience, and other factors.
3. The continuing need for information on resources to be provided to users as an essential function of network operation.[1]

It is reasonable to assume that a significant financial input to the start-up costs of a network would come from an appropriate government agency; that as part of this process a network planning team would be established; that many of the resources initially incorporated into the network would be university-based; and that the initial operational emphasis would be on research and education. Potential resources include hardware, software, instructional programs, other higher-level service capabilities, data bases, and people. Some steps a network planning team might take for investigating initial resources are

1. Make a broad nationwide inventory of the most obvious potentially available resources (drawing on existing partial studies where appropriate) to the limits permitted by money and time.

2. Evaluate the findings from the standpoint of scarcity, duplication, real availability as limited by projected commitments, possible legal constraints on shared use, and potential for meeting tentative standards for reliability, compatibility, and documentation. This step would have to recognize that the lead time between the initial inventory and actual network operation requires that good projections be made of the further development of identified resources as well as of presently nonexistent resources that may be available later.

3. Solicit statements of need from potential network users by well-designed survey techniques. The evaluated description of resources would be used to stimulate user perception of possibilities and even to stimulate the development of the initial market. The planner should recognize that although the survey of needs would show many areas that cannot be satisfied by immediately available resources, such needs must be considered in the initial decision process.

4. Evaluate the matches between needs and resources and between needs and funding sources to support these needs.

Standards and Quality Control

Critical areas of management concern in any network operation are the setting of standards for compatibility and the adoption of measures to ensure a high level of performance of products and services. An interesting question in network operation is whether refereeing and other evaluative techniques can be used as an alternative or supplement to fixed standards and quality control measures. The degree of control desirable is a related important question.

It seems that standards of various types will be necessary. Some must be set from the beginning; but others that may be desirable in the future may not be practical in the start-up phase since many initial resources will be selected from what currently exists. In particular, rigidly adhered to standards at the outset can inhibit important innovation, development, and use of network resources. In designing performance standards, currently approved standards should be used as guidelines unless there are good reasons to do otherwise, and they should be reexamined as changes are contemplated.

To protect network users and network investments, assurance of continuing high performance in the delivery of network resources is necessary. Measures of performance, both qualitative and quantitative, must be developed and continually reviewed. Some factors to be considered in developing such measures are

- Economy, cost benefit, and cost trend.
- Reliability of service, both short-term and longer-term.
- Comprehensibility, completeness, accuracy, and timeliness of documentation.
- Adequacy of user training, where needed.
- Match of service or program claims and results.

Network management should see to it that refereeing and quality assurance techniques are used continually to measure actual performance against the standards. Management should establish a forum for users to communicate about resource use among themselves and with network management and resource suppliers. It should undertake evaluative studies of network resources on its own initiative and make the results of these studies available to users on a continuing basis. The evaluative studies should be developed as a necessary part of a well-run service operation but should not be used to discourage offerings of difficult-to-evaluate services.

The problem of quality control in the provision of computing and data services is not new. It is chronic and perplexing, and it is not susceptible of easy solutions. Networking will exacerbate rather than lessen the problem as users become further removed from the source and developer of the programs and data bases they use. Both formal and informal mechanisms for evaluation and certification will need to be established. The assistance and support of the American Federation of Information Processing Societies, its member societies, and other relevant organizations and activities at the state and federal levels should be enlisted in this endeavor. Certain activities of the Atomic Energy Commission and the National Bureau of Standards come to mind in this connection, as does the statewide work, for example, of the University of Minnesota Committee on Statistical Packages and Support Programs.

Marketing and the Private Sector
Even the best-conceived network may fail unless an intensive and com-

prehensive marketing function is included in the operation and is pursued aggressively. To be successful, a network with a large number of heterogeneous users and demands calls for the application of marketing techniques that may seem unconventional in traditional academic circles, although they are familiar in the distribution of books and other educational and research materials.

Marketing techniques currently used by regional networks, textbook publishers, other business organizations, and government bodies should be carefully examined to learn what methods would be likely to be most effective in large-scale networking. Different user types will require different marketing techniques. Professional societies may hold a key to widespread marketing among certain discipline-oriented groups, for example.

Crucial to success will be the establishment of a *modus vivendi* with the private sector. The relationship must be one of symbiosis, not warfare. A national network cannot serve its users with one hand while using the other hand to fight commercial firms. Appropriate roles for commercial firms must be found in documentation and support as well as in marketing and distribution. The relationship between profit and nonprofit elements of the network should be made mutually supportive and complementary rather than competitive wherever it is to the advantage of users.

Providing a subsidy within research and development contracts for the maintenance, documentation, and marketing of a product could stimulate the interest of the private sector in playing an active role. Similar to "page charges" in the publication of technical articles, this subsidy, if instituted on a regular basis, would assure commercial firms of a sustained market for their services.

Effect of Different User Types
It is recognized that the requirements for the various functions that have been discussed—product development, documentation, quality control, and marketing—will vary with the types of user, use, and program or data base being used. The sociologist, inexperienced in computer use, who is working with classifications and averaging of simple survey data, has very different requirements from the experienced econometrician who is working with extremely large and complex data bases in modeling the United States economy. The sociologist will settle for the quality control enforced by the data collection agency and some heavily used canned statistical routines made available through the network program library,

with documentation of the cookbook variety. The econometrician, by contrast, will need access via the network to a variety of national surveys and census studies. He will need to know in some detail how the data were collected, what significance they have, and the likely sources and magnitudes of error. The econometrician will also use canned statistical routines, but he will also want sophisticated programming languages and other tools to support his modeling activity. He will want to know some of the mathematical theory behind the tools in enough detail to allow him to assess their appropriateness for his purposes and to suggest departures and variations that he might develop. He will take responsibility for the quality control of his own creations.

The network planning committee must take account of the wide range of different requirements that users will have and see that the operation of the network accommodates this variety. The variety will include

- Types of user: discipline, institution, job classification, geographical location, degree of computational expertise.
- Types of use: instructional, research-oriented, administrative.
- Kinds of programs: number crunching, natural language processing, specialized graphics, text processing and editing, information retrieval, simulation and modeling.
- Data bases: administrative, physical science, behavioral science, humanities, life sciences.

Governance and Allocation

The governance of a network can range from the laissez-faire to the completely directive. The alternatives vary from allocation and direction by a user group (on the laissez-faire end) to a strong, small, central governing board appointed by an appropriate authority or elected by the users. Midway between these extremes is the possibility of a two-house governing structure: one house consisting of an executive board that sets policy; and the other (lower) house consisting of a user group that has the right of recommendation for approval to the executive board, the right of veto on certain limited ranges of action of the executive board, and the right of overriding the veto of the executive board on certain decisions. A "constitution" with amendment power should be developed to define the divisions of responsibility and authority.

Assuming the two-house structure, a user priority algorithm could be determined by the governing board and ratified by the user group. In defining levels of importance of use, the algorithm should differ from field to field and should distinguish between the activities of research and teaching; but it should not operate to the disadvantage of those less able to pay where this can be avoided. That is, it should not be based only on ability to pay but should also recognize efforts for the general good.

A mechanism is required for allocating resources for both assigned (or contracted) services and for as-needed services on demand. This mechanism should ensure that no one user could preempt the entire service at any time.

Other Policy Issues
A central question is how the network operation is to be supported. Who pays—the federal government, states, universities, users? How much by each?

There must be an exchange rate that is mutually beneficial to the nodes of the network. This rate will be the wholesale rate, and each node will resell the purchased services to its users at a higher retail rate. The overall wholesale funds collected should exceed actual operating costs and the excess administered by the governing board, possibly to help subsidize the more costly services for users with the least funds. This is one way of treating the problem of the rich versus the poor or the privileged versus the underprivileged. Other methods should also be explored.

Seed funding seems essential from a variety of sources such as federal and state governments, private foundations, or a consortium of users. Individual users themselves should be able to acquire small grants from private or governmental foundations. Seed funding will be necessary for a number of years, but the network should ultimately be self-supporting, except that some form of subsidy may continue indefinitely to finance new projects, unless this can be accomplished by a fund built up from user revenues.

Fertile avenues of cooperation between the network and cable TV, microwave, and satellite services seem possible. The relatively high cost of communications relative to computing, both at present and projected into the future, and the current technical inadequacies for computing of communications technology pose problems for networking that warrant thoughtful consideration. Cable TV and satellite communication may be able to help as they are more fully developed and deployed.

Finally, there are legal issues having to do with the protection of privacy and restrictions on the "resale of communications" that could place critical constraints on how networking can develop. The advice of legal counsel and regulatory authorities should be sought on these matters.

Reference

1. Douglas C. Engelbart, "Coordinated Information Services for a Discipline- or Mission-Oriented Community," ARC Report, Augmentation Research Center, Stanford Research Institute, Menlo Park, Calif., December 12, 1972.

40

Network Economics and Funding

Report of Workshop 12

WILLIAM F. MASSY
Stanford University
Faculty Discussion Leader

Contributors:

SANFORD BERG,
University of Florida

RICHARD H. BOLT,
Bolt Beranek and Newman Inc.

GARY FROMM,
Data Resources, Inc.

JOHN HAMBLEN,
University of Missouri, Rolla

BERTRAM HERZOG,
University of Michigan

LAWRENCE G. LIVINGSTON,
Council of Library Resources

JOHN F. LUBIN,
University of Pennsylvania

JEFFREY NORTON,
Jeffrey Norton Publishers, Inc.

RANDALL M. WHALEY,
University City Science Center

JOE B. WYATT,
Harvard University

MARSHALL YOVITS,
Ohio State University

Introduction

The purpose of this workshop was to examine those aspects of network-ing that relate to economic organization and feasibility and to planning. We were concerned with a variety of issues. For example, one of the par-ticipants outlined the following question in a written communication before the seminar:

What—in very specific and mundane terms—is any network to accom-plish, and what will it cost, and what will be its cost effectiveness in comparison with alternate ways of accomplishing the networking objec-tives?

Inevitably, the development of a network will be based on a series of assumptions. All too frequently much of the effort thereafter goes toward justifying the initial premises. Such tendencies should be guarded against by making sure from the outset that there is a completely objective evalu-ative body that is constantly monitoring the *relative cost* and the *relative usefulness* of the information provided in the network, most par-ticularly as compared with alternate available methods for acquiring the same information. Experience with users cannot be the only data cited; analysis of the needs of *nonusers*—and why they are nonusers—must also be included.

In considering economics, the cost of alternate nonnetwork solutions should also be examined. For example, for a given body of data, could multiple local resources and local accessibility lead to lower costs and greater use? Also required are much-improved techniques for determin-ing cost effectiveness of various information sources (a major concern for libraries as well as for networks).

It would also be desirable to determine whether certain classes of in-formation are suitable for network access and some are not. Such a determination could make a major difference in the funding require-ments.

Finally, any networking system should be funded by making grants di-rectly to the user rather than to network operators. This arrangement would make the information consumer the ultimate determiner of what is provided in the network and would provide a choice among a variety of commercially available services.

As this presentation shows, the matter of network economics and funding involves a good deal more than simple questions of dollars. One proposed taxonomy of issues included: (1) users and types of users, (2) the values and cost-benefit trade-off involved in networks, (3) the organi-zational reality of network usage, and (4) how the forces underlying the first three areas would evolve over the next ten years. Needless to say, the workshop was able to address only a limited number of issues during the time available and to address these in only a general way.

Members of the workshop divided into three subgroups whose topics form the three major headings in the remainder of this report. They are (1) market structures and regulatory considerations, (2) planning for network development, and (3) key needs in network research.

Market Structures and Regulatory Considerations
The objectives of this subgroup were to examine some questions relating to the market structure and regulatory needs of a communication network consisting of a number of independent or quasi-independent computing units linked together by a communication system. The communications system is viewed as providing a market for computing services much as an organized stock exchange makes a market for individual traders in securities.

The following selection from a paper by one of the participants reflects the flavor of the workshop's view of such a market. The conceptual structure is from the field of economics, and the context is the status and prospects for breaking down the tendency toward monopoly in large-scale on-campus computing activities.[1]

Without wishing to force the trade analogy, the current absence of computer networks in the economics discipline seems to be a perfect example of autarchy: many nations (universities) trying to be completely self-sufficient in the production of goods (knowledge within particular disciplines)* as though transportation costs were so great that trade between nations were not feasible. Introduce to this concept the analogue to speedy, low-cost shipping: computer networking. Universities, like nations, could mistakenly react by setting up tariffs to reduce the volume and avoid dislocations caused by instantaneous free trade.†

In addition, university administrators would resist what might be perceived as "colonial exploitation." A department may feel threatened at not having its own in-house computing facility. It may believe itself doomed to remain on the periphery of methodological developments in the field if another university in the region has all the supporting personnel and data on its premises. Designers of any computing network should

---

* Of course, researchers do use the output of research conducted elsewhere in the production of their own research. Science, after all, is a cumulative process. That production may not be as efficient as it could be, however, if a further division of labor could be devised. Note that for a few researchers the benefits of working with scientists at other universities or using systems developed elsewhere outweigh the costs of doing so.
† Fortunately (as well as realistically) universities do not have to fear having to cope with such a radical change overnight. Even if the network were to appear tomorrow, researchers would not be able to utilize it. It might take a decade before it became a standard research tool.

anticipate such reservations within departments. The theory that "trade follows the flag" (and vice versa) conjures up visions of foreign domination resulting from trade. If departmental fiefdoms will not cooperate, the university itself will have a difficult time agreeing on some resource-sharing arrangements with other universities.

An alternative theory of trade is more widely accepted today: the theory of comparative advantage. Basically it takes the resource base of nations as given and tries to explain what products they will export and import in terms of the inputs they have in relative abundance.

Let us take two products from economics to illustrate what the theory has to say. One field, consumer behavior, might require minimal computing power (in fact, the storage requirements for consumer studies may be substantial) but a substantial amount of personal services such as collection of survey data, data entry, correction, and counseling. The other field, simulation models of the economy, draws from a data bank already in existence but requires substantial computing power to facilitate the analysis of alternative specifications, estimations of effects of various policies, and simulations with different types of stochastic shocks. Assume that there are two universities, each of which attempts to remain self-sufficient. If one university already has the large computer and the other already has additional budget lines for personnel, there would be room for a deal that could make both departments better off. Institutionally speaking, this switch (sharing of resources) is easier than if one university agreed to handle "all" economics computing while the other handled "all" biology computing.

A number of conditions can invalidate the basic conclusion that movement toward "free trade" (and greater specialization within countries) expands the potential output of all nations together. First, there may be distortions in the existing pattern of prices. Some universities may have subsidized computer facilities and if this advantage is financial rather than reflecting real resources, then adjustments might be inefficient. Similarly, conditions may prevent the movement of personnel and computers to where they can be used most efficiently. For example, if a university has purchased a machine or has a long-term contract it will not find purchasing computer time from another university an attractive idea.

The noneconomist will more easily recognize the "infant industry" argument found in connection with trade policies of developing countries. These nations do not want to take the initial resource endowments as given; rather they wish to restructure their economies through export promotion or import substitution policies. Thus barriers may be raised to permit an industry to gain a foothold, when exposure to external competition would doom it. Of course some infant industries never grow up, so protection may generate no benefits other than those that accrue to the protected industry. Resources are not released for the development of other sectors.

A key lesson to be learned from trade theory is that the form of protection affects efficiency. For example, a university that fears for the financial viability of its computer operations might attempt to put quotas on the use of outside computing facilities. Such a policy has some very detrimental side effects. First, who is to be allocated the outside time? Whichever researchers are granted the quota have an advantage over other researchers from that department. If a first-come–first-served policy prevails, then a rush to use the super foreign facility will occur, with obvious inefficiencies accompanying the artificial speedup in activity. The same limitation on foreign computing could be attained through a tariff (or tax) on outside usage. The users who place the highest value on the outside facilities will be willing to bear the higher real price. In addition, as demand for such computing grows, it will be supplied by the outside facility. If this threatens the financial viability of the home computer the tariff could be revised upwards, but at least the administration would have a gauge of the perceived value of the outside facility.

The notion of financial viability brings up the problem of underemployment of the home facility. This situation is analogous to attempts of governments to cut back imports (through higher tariffs) to expand employment at home in a self-defeating "beggar my neighbor" policy. When other nations retaliate, employment in export industries falls, leaving everyone worse off. Nations pay higher prices for products, and overall employment is not improved. Clearly a better policy would involve using internal policies to expand utilization of existing resources.

International economics can also suggest techniques and institutions for facilitating the adjustment process. The formation of regional customs unions is not as threatening as the establishment of free trade with all nations, particularly if the production structures of the participants are complementary rather than competitive. There may be trade diversion from countries outside the customs union, but there is also trade creation and improved division of labor within the common market. There are also a number of dynamic advantages from such arrangements. Thus regional consortia within economics (and other disciplines) may prove to be one way of moving to more efficient production of scientific knowledge.

In addition, some national cooperative arrangements may prove to be useful for cases when all participants understand the advantages of specialization in production. GATT (General Agreements on Tariffs and Trade) sets institutional rules limiting artificial barriers, and a similar institution would help delineate proper and improper (inefficient) ways of protecting departments. The basically liberalizing influence would be useful. Then too, an institution corresponding to the International Monetary Fund would have a role to play. That is, some units of account need to be developed so that trade can grow efficiently. Since interinstitutional payments would put a strain on finances, some accounting procedure could be developed that would facilitate multilateral rather than bilateral balancing of credits and debits. Perhaps a university undergoing a severe adjustment would be permitted to borrow for a period of time.

Three problems (again analogous to those in balance of payments) arise: adequacy of existing interinstitutional arrangements, ways to facili-

tate the future growth of transactions, and ways to facilitate the adjustment process. As key personnel shift affiliations, as computer technology
changes the economics of network requirements, universities will need
to be able to adjust and make the proper investments in technologies
and personnel. One should not underestimate these problems, but
neither should they prevent "institution-building" in this area.*

The reader will recognize the reemergence of many of the ideas discussed in organizational and political terms in the write-up of Workshop
6, Chapter 27, on institutional relations.

A Market Structure for Networking in Higher Education

In order to examine some of the market implications of a distributed
or facilitating network, the subgroup developed a "paradigm," or description of the end point to which such a network might evolve. It is not
strictly predictive or prescriptive but rather is a combination of the
group's prognostication and hopes.

Suppose that a network for higher education consists of P nodes. Some
of these would be host nodes, in the sense of being potential suppliers
of computing service. Let us call these real hosts and assume that
there are m of them. The other $P - m$ nodes would not be potential suppliers of service. For purposes of the present discussion we shall assume
that some but not necessarily all the host nodes are colleges or universities. (Other kinds of organizations like nonprofit or profit-making
corporations and national centers can also be viewed as real hosts.) Similarly, some but not necessarily all the remaining $P - m$ nodes would be
colleges or universities. The communications network might be owned
or operated by a profit or nonprofit corporation, a consortium of uni-

---

* Note that we have focused on trade *theory* to determine commercial *policy*. Constraints and deviations from assumptions complicate the policy issues. Also, the entire
"balance of payments" of a particular computer center could have a deficit on the
trade of computer services. Some universities may find it cheaper to have no in-house
computing capability, since access to other facilities may be more economic given the
structure of demand. Some National Science Foundation "Special Drawing Rights"
would smooth the adjustment process and provide incentives for more efficient use of
our computer resources.

If universities did want to balance trade, a "floating exchange rate" would be the
theoretical ideal, permitting all computers to be used to capacity. If one center became
overloaded with demand, prices there should rise, causing the amount of computer
time demanded to fall. Users are responsive to price and would substitute more labor-
intensive processes for computer time. Eventually the higher prices would justify additional investment in facilities. Determination of the location of such facilities is no
simple matter.

versities, government, and so forth. The present model assumes, however, that the communications network is a common carrier with respect to its dealings with individual nodes.

In addition to the P nodes on the network, we assume that there are Q special service organizations such as consortia of universities or professional societies or academies. Their purpose would be to stimulate and possibly arrange for the providing of special kinds of computing services like particular data bases or software systems. A special service organization might contract with one or more host nodes to supply the service on the network, or it might content itself with setting standards, trading data or programs by mail, and the like. The special service organization can be thought of as a pseudonode on the physical network or, perhaps better, as a node on an extended logical network encompassing but not limited to the physical network.

It is also useful to distinguish between users, retailers, and wholesalers of computing services. Users are defined as individual members of universities or other organizations with access to the network. Users do business with retailers, who may be organized around a university or special services organization. Wholesalers are owners of a host computer center or of a special service organization (perhaps involving a proprietary data base or software system). Retailers do business with wholesalers to obtain goods to be delivered to users.

The essential difference between a retailer and a wholesaler is that the former is a full service organization in the sense of providing complete documentation, user assistance, and other services as well as computing power, data bases, and programs as such. Wholesalers, on the other hand, provide machine cycles of various types in various configurations, access to data bases, and so on. It is clear that an individual computing center can function as both a wholesaler and a retailer. The important distinction is that wholesalers will be expected to charge lower prices per computing cycle or data access than a retailer would for the same service. For example, a campus computing center might charge a retail rate to its users (thus accepting the responsibility for providing user services) and a lower wholesale rate to other retailers on the network. The wholesale computing service might be limited to raw machine cycles. But it could also include access to data or programs with documentation supplied on an "as is" basis, with specified consulting services available only to retailers' professional personnel.

A college or university computer center would have to accept the responsibility of being a retailer to its users, whether or not it obtained its computing in-house or from the network. If computing is obtained from a foreign host, the retailer would be expected to mark up his cost to compensate for the expenses of providing user services at retail. Furthermore, any university would be free to limit the retailers from which its own users could obtain computing services. For example, it would be possible (though not necessarily desirable) to require all users at a given university to obtain services from the local "computation center" whether it owns a computer or is merely retailing computing services.

Each retailer is a node on the network dedicated to supplying a group of users that is either localized geographically or bound together by some other relationship. Examples would be members of a given university community or researchers authorized to use a specific discipline-oriented center. Each retailer is viewed as setting his own prices for computing services.

It is possible, even likely, that regulation of the wholesale price of computing services will be necessary. The reason is that different organizations will have different accounting methods for capital costs, equipment amortization, and operating expenses. Lack of common accounting standards might cause a chaotic wholesale price situation.

The last condition in our hypothetical market structure is that of freedom of entry on the network for new nodes, both wholesale and retail. New retail nodes do not pose a particular problem, since they would be expected to pay the going rate for wholesale services plus the cost of communications services. The situation is slightly more complicated for a new wholesaler. Two methods of "buying in" are possible, depending on the ownership of the facilitating (communications) network. If the network is owned separately, the new wholesaler might be asked to pay the cost of connection to the network plus some kind of "marketing and accounting" fee. If the network is cooperatively owned, perhaps by a consortium of wholesalers and retailers, a new wholesaler might be expected to buy into the group. In this case it would be purchasing some of the equity of the consortium. In both cases the prospective wholesaler would probably be expected to sign a contract requiring certain standards of service and incorporating an agreement as to technical and accounting standards.

While this description of a distributive network's market structure is quite sketchy, the workshop believed that it had certain important elements of reality. In particular, it provides for a highly flexible and economically viable method of market organization. Further work in refining these or similar concepts would be valuable.

Regulatory Scenarios

A number of regulatory scenarios are possible, depending on technological, behavioral, or institutional constraints. These constraints correspond to the three traditional justifications for regulation: natural monopoly, destructive competition, or an overriding public or social interest (externalities).

In the first case, if the scale economies are significant compared to the size of the market, a single seller may be most efficient, but he would have to be regulated to eliminate monopoly profits. The signal transmitter as a common carrier would be an example of such a situation. Note that this is a technological situation that could change with the advent of new technologies. Sometimes "system integrity" is offered as a justification for operation by a single entity.

Destructive competition may be said to occur when entry into an industry is easy but exit is difficult. The situation can result in short-run price wars, bankruptcies, deterioration of service, mergers, monopoly profits, and then a new cycle. Such fluctuations are not likely to serve a useful resource allocation function and will result in pressure for regulation. Again, technological change may change the underlying conditions. If university computer centers just entering a network *all* offered very low prices (due to short-run excess capacity), destructive competition could result if users were highly responsive to price (as opposed to other service variables) and there were no learning costs or inertia.

The social benefits result either from increased benefits or from decreased costs owing to regulation rather than private decision making. These benefits could be lumped under "intangibles," or one could attempt to quantify them. Loss of control, premature disclosure, altered incentives for innovation, or the loss of a joint product would have to be taken into account when evaluating the effect of the "loss" of a physical computer center. If a natural monopoly is involved, the gains of unitary management would have to be balanced against any negative externalities.

An additional justification for community intervention (if not regulation) is that the good involved is a public good, like national defense, or some types of information. For such goods, more for one party does not imply that there is less for another, and so the provision of the optimal level of such services involves collective action. The establishment of programming systems or data bases certainly shows this characteristic. But exclusion is possible for these public goods—that is, potential consumers can be prevented from gaining access to proprietary data bases or software systems. Even though their valuation of the product is greater than the incremental cost of providing it to them, an even higher price might be set to recover the fixed costs of creating the data base or software. This possibility might even be efficient as an incentive to produce more of such services. A certain amount of pluralism is highly desirable. Multiple centers of initiative are more likely to achieve true breakthroughs than a regulated monopolist.

Planning for a Network

The Planning Process

Planning for a resource-sharing network should be a continuing process involving preplanning to establish basic concepts, detailed planning of the steps from start-up to stable operations, and a procedure to ensure that the system remains responsive to changing conditions, opportunities, and the needs of users.

The initial plan should serve several purposes. It must be complete and substantial enough to serve as a basis for sound decisions. The directions or actions it recommends should be clear and should be justified by reference to the alternatives that were considered. It should provide a logical basis for solid development of the system. If the decision to establish a network has not been made, more preliminary work like feasibility analyses and weighing of political and economic factors will be necessary. The decision must be reached by the agency that has the authority and, in principle, the access to support that will make a decision to create a network workable.

The plan for resource sharing among institutions by means of a network must be comprehensible to different audiences, among them top administrators and decision makers. It must therefore avoid technical jargon. It must, however, be comprehensive enough to describe clearly all the technical issues and factors that are relevant to the development of a plan that is workable and can be implemented.

Terms need to be defined. Even the workshop, for example, had difficulty deciding on a definition for a network. The very word computing was found to have different meanings; some use it in a broad generic sense while others limit it to number crunching.

The plan should contain procedures for evaluation at specific stages of the network's development to allow for refinements based on experi-

ence, to make sure the system is cost effective, and to allow changes
in the scope or rate of development if necessary.

The following pages contain suggested sections for the plan.

A Development Plan for a Resource-Sharing Network

Summary
Written as a stand-alone section. Avoids technical jargon, but complete
in all essential elements for indicating with clarity the objectives,
nature of operations, and benefits to be derived. Aimed primarily at top
management of institutions and organizations to be tied into the network.
Cross-referenced to later sections of the plan.

Introduction
Contains brief historical matter putting the plan into the context of
other national developments in computing. Indicates what actions led
to development of the plan and the decisions, and by whom, that make
worthwhile the work toward implementation of a network. Reference
should be made to relevant data, studies, publications.

As needed, terms should be defined in this section.

The sequence of developments should be reviewed, indicating the
content and the nature of the sections to follow.

General Requirements for Computing
This section provides the various elements of justification for the decision
to develop a resource-sharing computing network. This very important
review should persuade the reader that the need is real and that the
various requirements were considered in arriving at the specific operational objectives set for the network. Elements or factors include rising
costs, increasing demands in quantity and diversity for computing,
desires to share special programs and applications, need to make provision for extending valuable regional services and programs to other
parts of the country, and so on.

Specific Objectives for the Network
This section of the plan should focus on the basic elements of the
proposed network, clearly indicating the measurable objectives and functions to be served. Subsections might be ordered as follows:

1. Definition and justification of the specific network recommended.

- Reasons for choice of system as providing capability for effective sharing of computing resources.
  2. Functions to be supported or to be performed by the network.
  3. Prospective membership of the network, institutional or regional centers.
- Selection procedures and reasons.
- Relations between "members" and ultimate users of computing.
- Prerogatives and responsibilities of membership.
- Contractual relationships.
  4. Hardware.
- What hardware will be necessary to operate the communications network.
- What additional hardware will be necessary to interface users and resources.
  5. Communications.
- Technical characteristics required to perform functions and meet objectives.
- Alternative communications systems.
- Characteristics of the communications system; for example, main trunks with major nodes and regional nets.
- Means of gaining access to system; whether owned or leased.
  6. Network management. Management functions include all aspects appropriate to corporate operations:
- Production and delivery of services.
- Marketing and relations with customers.
- Realistic pricing to establish fiscal stability.
- Financial management, accounting, and so forth.
- Support of consultative and user education and documentation services.
- Review and evaluation of all operations, continual planning.
- Authority to negotiate contracts within established policy.
- Personnel management.
  7. Network corporate organization.
- Structure of organization; composition and authority of board of directors; means for providing technical advice at the highest level.
- Legal character of the organization; profit or nonprofit; related to or independent of government agencies; corporate role of "members."
- Headquarters location and facilities as required.
- Procedures by which administrative staff will relate to the board and will be responsive to both the needs of users as they emerge and the general objectives set by the board.

Funding

Specific attention to funding during three phases of network development and operation:

• *Start-up.* Costs associated with implementation of all phases of the plan, encompassing personnel, expenses of operating a network office, accumulation of information, communication among regional and institutional centers, and acquisition of capital goods and supplies.

• *Network capitalization.* Possibly considered to include both the costs of acquisition by purchase or lease of network-related hardware and the establishment of a working-capital fund to enable the network to operate while receivables grow.

• *Network operations.* Major revenues to come from realistic charges to users; support from public funds or quasi-public sources considered to be in the public interest at large, not appropriate to be passed on in distributed charges to users; government contracts and grants not considered as subsidies if based on specific contractual obligations; institutional or regional computing centers consider the incremental costs of computing associated with network charges to be continuing standard components of computing costs to the ultimate user, more than offset by economies of scale.

Plan for Implementation

The process by which a decision is reached regarding a resource-sharing computing network is inevitably an iterative process. Repeated passages through the elements of a plan will occur before major and subordinate decisions are reached.

In this section steps are described that have to be taken once a firm decision is reached on the organization and initial funding of a network. It is important to recognize that any steps toward implementation will be fruitless unless firm commitments from one or more sources of funding are assured. It is not essential, however, that funding for all perceived steps be assured from the outset. It may well be possible to assure funding only up to a planned checkpoint for review and evaluation. Even so, it would be highly desirable to have contingent commitments for continuation of funding, with the decision to rest on results of the initial phase.

The implementation steps should lead logically from one to the next, or in some fashion as appropraite of a PERT (Project Evaluation and Review Technique) or CPM (Critical Path Method) plan of action. It should not be left to chance at some later date to try to connect into a plan a series of uncoordinated and independent studies and research projects. These have value in early stages of consideration of whether networking is potentially beneficial and feasible. There would be merit, however, even in conception of studies and research projects, which may still be needed, to have them considered for value as steps toward implementation of a coherent network plan of operation.

The Long-Term Future
Although it would be premature to project details for further developments until a network is operating and stable, there would be merit in identifying steps to be considered.
- Interconnections with other networks.
- Changes based on advances in communications and computing technology.
- Changes in priorities in use of public funds for services (in education and health, for example) and the potential of the network to provide certain services on a cost effective basis.
- Conversely, with growth of service in magnitude and in scope, changes restricting or limiting use of the network to avoid hazards of complexity and confusion from unplanned growth and scope.

Key Research Needs

On the Need for Assessing the Potential Value of Networks
What do we mean in this paper when we use the word network? In physical terms, we mean a set of interconnected computer facilities in which at least two of the computers or terminals are farther apart than the average dimension of a campus. In functional terms, we mean a system that can replace or improve on some or all of the capabilities supplied by a local computer center. Further, we restrict our attention to networks for serving research and teaching in higher education, and research in science and engineering generally. Such activities, according to anecdotal evidence, could receive such benefits as lower cost, greater reliability, increased versatility, and perhaps entirely new opportunities.

During the past decade, several networks of different kinds have come into existence and are serving an expanding number of uses and users. The experiences gained have generated considerable enthusiasm and numerous proposals for new network developments. Yet up to the present, the justifications given for moving vigorously into the world of networks appear to be based almost entirely on faith and assumptions. Almost totally lacking is such empirical evidence as statistically valid measurements and systematic observations on the performance and utilization of the networks that already exist. Such evidence would appear to offer an important, if not essential, underpinning for responsible statements concerning the likely advantages and disadvantages of networks in the future. Without such evidence, predictions of relative costs and benefits of networks versus local facilities can be little more than guesses.

Clearly, a network is not just one thing. Subsumed under the definition we have given could be different networks serving quite different functions and taking quite different forms. And several networks of different function and form could be combined as subsystems of broad-scale network systems. What are the more realistic alternatives as to the function and form of networks and subnetworks? Within each common category, what are the benefits and costs of definable alternatives? How do these costs and benefits relate to those of nonnetworks that do the same jobs? Upon what criteria can wise choices be made among alternatives? How should the chosen alternatives be started? And, over the longer haul, how should they be operated and managed?

These are the broad questions that must be answered. Developing credible answers will require intensive analyses of technological, economic, sociological, financial, and other aspects. These analyses, in turn, will be more realistic the more they draw on empirical evidence gleaned from operations that already exist.

Figure 40.1 illustrates how a set of investigations might be organized. This section outlines the broad questions and their related issues in sufficient detail to express what we have in mind. The paper then suggests several kinds of investigative evidence that could be gleaned from

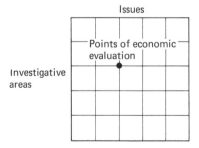

Figure 40.1. Evaluative model structure.

existing network activities and would serve to clarify the issues and enrich the analyses upon which the issues will be resolved.

Types of Investigative Evidence

One possible grouping of the investigative evidence to illuminate the broader issues is as follows:

1. A taxonomy of user services to be offered by the network.
2. A taxonomy of resource-oriented services to be offered by the network.
3. A set of information communication capabilities that would provide to multiple classes of users a set of consultation and training modules related to the services available and the use of the network and the related resources.
4. A set of distribution resources to connect different types of terminals to different communication interfaces for users.
5. A comprehensive set of cost-accounting and billing capabilities to recover costs fairly and to provide data for evaluation of costs and benefits.

For each of the categories of investigative evidence, the value added and the cost should be found and evaluated so that a price structure can be developed. For future planning and evaluation, an economic projection of the technological and other phenomena affecting the effective "balance" between network services and resource services should be included relative to each of the basic questions.

Key Issues in Networking

It is next appropriate to examine the second dimension of the model, the basic questions previously described. Each of the questions raised within the five categories of issues represent a subject that would be evaluated relative to the categories of investigative evidence.

WHAT IS A NETWORK? Can we use as a test the notion that the service must be better (at least equal to) than presently available to the average user? Are not a number of definitions of network needed, and thus will it not be necessary to consider alternative futures for each kind of network? What is the effect of considering a network as a national or even global entity? Is not the notion of regional networks more practical at this time? Discipline-oriented networks? Is it possible to make a statement of objectives for a network? What should it be? Is it not possible that the real advantage of a network is that it can provide access to different kinds of capability, not cheaper or better versions of the same capability? Does it makes sense to think of one network? Will not the final network inevitably be a network of networks?

WHAT CAN A NETWORK DO THAT IS BETTER THAN OTHER POSSIBLE STRUC-
TURES? What tests can be established to show whether a network is better
or not?

What will be the effect of technological change on this issue? If, for
example, the cost performance of minicomputers increases and CPUs
and memories drop radically in price, are any of the current notions of
"network" viable for the next five to ten years? For this kind of study,
what is a proper definition of cost? Benefits? Value added? Quantification
of intangibles? What alternatives are there to networks anyway? Is
the following true: "Building a network will cost more than that existing
before: it cannot be justified on quantifiable factors alone but on intangibles"? What are the benefits of networks? Do they: Equalize opportunity of access? Lower capital and operating expenses? Mean more
computing power delivered to users? Save user time (increase user productivity)? Enable good people to go anywhere?

What are the disadvantages of networks? Do they: Increase academic
elitism? Mean a loss of control for university administration? Result in
poorer service? Imply a loss of prestige for some institutions? Tie an
institution into an inflexible arrangement? Lead to a loss of safety, security, or privacy?

Is a network better? Would we pay more for overhead, broadly defined,
in any network than can be compensated for by any possible economies
of scale?

It seems clear that conventional economics and economic reasoning
are inadequate to consider the cost and benefit of networks. What
analytical mechanisms can be developed to measure the economic and
organizational realities of networks?

WHAT FORM OF "NETWORK" IS BEST? Suggestion: Clinical studies of existing networks to try to develop generalizations that might help in determining the best form of network. Is it possible to design an optimal
network? Is the notion meaningful? If not, what?

HOW SHOULD A NETWORK BE ESTABLISHED? What about funding? Organization? Personnel? Equipment? Relationship with common carriers?
Relationship with federal, state, local agencies? Should such a network
be limited to higher education? All education? Commercial users?
Commercial suppliers? How best can institutional, organizational, and
personal obstacles to establishing a network be overcome? How best
can a network be established, by evolution or by fiat? Out of task-oriented
networks? By establishing a national, full-function network de facto?

What are the advantages and disadvantages of alternative financial organizations? What capital requirements will be needed in the organizational and start-up phases? What will be the sources of that capital?
HOW SHOULD A NETWORK BE OPERATED AND MANAGED? What characteristics
are needed by the management of a network? What management
principles apply to operating a network? What about pricing? What are
the economics of the wholesaling-retailing notion? Should agencies
subsidize suppliers or subsidize users? How would it be best to supply
ancillary services, such as consulting, documentation, and maintenance?
How can protocols best be developed? Should entry be restricted? Should
networks exert control over substance or be facilitators only? What are
the advantages and disadvantages of alternative methods of recovering
costs from users? How can economic incentives be used to help balance
supply and demand in networks in both the short and the long run?
What will the capital requirements be for operation of the network?
What will be the sources of that capital?

Reference

1. Excerpted from Sanford V. Berg, "Networks in Economics," *Networks and Disciplines,* Proceedings of the EDUCOM Fall 1972 Conference, EDUCOM, Princeton,
N.J., pp. 28–35.

Index